ETI: A Challenge for Change

ETI:
A Challenge for Change

Peter Schenkel

VANTAGE PRESS
New York / Los Angeles / Chicago

Contents

Abbreviations   ix
Preface   xi

Part One. The Global Dilemma

1. Our Puzzling Stalemate   3
2. Lost Faith in Politics and Politicians   6
3. The East-West Bottleneck   16
4. The Deepening North-South Gap   28
5. The Tragic Spiral of the Armament Race   41
6. World Peace: A Dream Gone Wrong   60
7. The Quicksands of Revolution   72
8. Science and Technology: Between Glorification and
    Damnation   86
9. The Escapist Urge   102
10. The Global Deadlock   117

Part Two. ETI and Our Stalemate

11. The Basic Proposition   135
12. Search for Extraterrestrial Intelligence: An Impressive
    Record   140
13. Probabilities that Extraterrestrial Life and Intelligence
    Exist: The Stand of Science   150
14. Number of Extraterrestrial Civilizations: Estimates and
    Debate   159
15. ETI: Its Advanced Technological and Organizational Order   170
16. Contact with ETI: What It Would Mean to Us   183
17. The Task Before Us   199
Epilogue   215

Appendixes

Appendix A   223
Appendix B   225
Appendix C   227
Appendix D   229

Notes   231

Index   243

Abbreviations

AAAS—American Association for the Advancement of Science
ABM—Anti-Ballistic Missile
ALCM—Air-Launched Cruise Missile
APRA—Alianza Popular Revolucionaria Americana
ARDE—Alianza Revolucionaria Democrática
ASAT—Anti-Satellite (weapons)
CEP—Circular Error Probability
CETI—Contact with Extraterrestrial Intelligence
COSPAR—International Committee on Space Research
ERP—Ejercito Revolucionario Popular
ETI—Extraterrestrial Intelligence
FARC—Fuerzas Armadas Revolucionarias de Colombia
FDN—Fuerzas Democráticas Nicaraguenses
FSLN—Frente Sandinista de Liberación Nacional
GNP—Gross National Product
IAA—International Academy of Astronautics
IAEA—International Astronomical Federation
IAU—International Astronomical Union
ICBM—Intercontinental Ballistic Missile
ILO—International Labor Organization
INF—Intermediate Range Nuclear Forces
IRAS—Infra-Red Astronomy Satellite
ISSOL—International Society for the Study of the Origins of Life
LDC—Least Developed Countries
LOT—Large Optical Telescope
MAD—Mutual Assured Destruction
MBFR—Mutual and Balanced Force Reduction
MCSA—Multi-Channel Spectrum Analyzer
MIR—Movimiento de Izquierda Revolucionaria
MIRV—Multiple Independently Targetable Re-entry Vehicle
NATO—North Atlantic Treaty Organization
NIEO—New International Economic Order
OAS—Organization of American States
OECD—Organization of Economic Development and Commerce
OLP—Organization for the Liberation of Palestinians
OTA—Office of Technology Assessment
RAF—Red Army Fraction
R & D—Research and Development
SALT—Strategic Arms Limitation Talks

SDI—Strategic Defense Initiative
SETI—Search for Extraterrestrial Intelligence
SIPRI—Stockholm International Peace Research Institute
SLBM—Submarine-Launched Ballistic Missile
SLCM—Submarine-Launched Cruise Missile
UNCST—United Nations Conference on Science and Technology for Development
UNCTAD—United Nations Conference on Trade and Development
UNDP—United Nations Development Program
UNIDO—United Nations Industrial Development Organization
WHO—World Health Organization

Preface

I don't know when I first began wondering about the possibility that intelligent life may exist elsewhere in the universe. As a boy, I would vainly look for the man on the moon that my mother kept pointing out to me on clear nights, when its silvery disk bathed our hilly vineyard country in northern Yugoslavia. Though I could never make him quite out, I would always feel awed because, as my mother used to say, he is a very wise man. And certainly, she would add with a stern tang in her voice, I could learn a lot from him, especially how to behave.

When I mused again about our likely brothers in space, it was much, much later. The horrors and hardships of World War II, which I somehow managed to survive, and the meager postwar years in Austrian exile, belonged to the past. But the new global confrontation of the late forties and early fifties had already demonstrated that the precious peace, obtained at such high costs, was built on shaky ground. One day, in Southern California, while picking oranges near Anaheim for fifteen cents a box, a queer idea occurred to me. I wondered whether on a faraway planet, other beings managed to settle their differences in a more gentlemanlike manner, affording a recent graduate of political science, as myself, a more befitting way for making a living.

Since then, my vague notions on extraterrestrial intelligence (ETI) have come into much sharper focus. What contributed most to this were the amazing scientific breakthroughs that have revolutionized man's understanding of how the universe works, and how life and Homo sapiens may have come into being. They have all but stripped *Camille Flammerion, Jules Verne, Aldous Huxley* and other distinguished forerunners of today's science fiction of the glamorous fantasy and utopian distinction which gave their work such an exciting quality. Existence of other intelligent life forms in the galaxy and the possibility of making contact with them is no longer considered a blasphemy or foolish daydreaming. Science has advanced compelling reasons for the assumption that life and intelligence are probably quite abundant in the universe and that many of the extraterrestrial civilizations are likely to possess a social and technological order much superior to our own.

The near certainty of other advanced worlds is a challenge of profound meaning for man at this particular crossroad of history. An excruciating global crisis besets our contemporary civilization. In less than a lifetime since the genocidal conflagration of World War II, man has managed to make a shambles of the precarious balance that emerged from its graveyards, and to play havoc with all hopes for a genuine and lasting peace. Apprehension that vicious circles of confrontation are pulling mankind deeper and deeper in the gorge of a relentless maelstrom is mushrooming everywhere.

Because of this gloomy state of our world affairs and their even bleaker perspectives, I can't think of a more stimulating and rewarding adventure than to make the possibility of man making contact with another intelligent species in space come true. I think that the discovery that we are not alone in the universe, that other fabulously advanced civilizations exist, would impact mankind like a true day of awakening. It would strengthen our badly crippled common bonds and help set off a catharsislike metamorphosis, enabling man to eventually cut through the Gordian knot, in which his petty problems and quarrels are woven. The clear perception of his cosmic dimension may spur a new sense of responsibility toward the human race as a whole and endow him with the wisdom for guiding the adrift vessel of our civilization in quieter and happier waters.

Though search of ETI has become a solid scientific proposition, in many quarters it still remains a suspect subject. This curious fact reminds me of an incident, back in 1957, when the Soviet launch of *Sputnik I* startled the world. When I mentioned the news to an acquaintance in Veracruz, he pointed up to the sky and said disparagingly, "I don't believe it's there. I don't care what everybody says. It's just propaganda!" With regard to ETI, many people still react in the same way. It is my aim to dispel this skepticism and to show that there is much more to the issue of extraterrestrial intelligence than the UFO hoax and the fanciful concoctions of *Erich von Däniken*.

In Part 1, which deals with the crisis syndrome of our time, I have been able to take advantage of my own personal experiences and observations. World War II has been an excellent teacher against war and violence. During the subsequent years as a student, I learned that hunger is a rough customer. Much later, while doing research and managing development projects in Latin America, I came to grips with structural poverty and underdevelopment and with the forces resisting long overdue transformations and social change. But the years in Cuba, in the early sixties, and later in Chile, during *President Allende's* shortlived experiment, also taught me that revolutionary zeal and cocky dogmatism are poor advisers and that they do more harm than good. For over four decades I have been a frustrated bystander of the appalling downfall of statemanship on the world stage, and a witness of the grisly East-West confrontation menacing man's survival today.

In Part 2, dedicated to the central proposition on ETI, I have drawn on many sources, to all of which I am deeply indebted. *Carl Sagan's* marvelous explorations of the subject have been a fountain of constant inspiration. So were *Isaac Asimov's* brilliant insights, and *Frank Drake's* pioneering search experiments and messages into space, which converted SETI into the fascinating scientific enterprise that it is today. The outstanding work of *Michael D. Papagiannis,* first president of the IAU Commission 51 "Search for Extraterrestrial Life" has also greatly enriched my views, as have the contributions of many other brilliant scientists. What I owe to them cannot be expressed in simple words. They are entitled to share with me any favorable comment which this work may merit. For the omissions and mistakes, I alone am responsible.

My deepest thanks belong to my wife, Margarita, for her incommensurate patience and indulgence with me during the years I spent writing and rewriting this book. I don't know whether I could have finished it without her constant care and loving understanding. I hope she and our three children, Frieden, Sianna-Amistad, and Aimée-Niqué, can forgive me for depriving them of my company during the short spells of time that I was able to devote to this task.

ETI: A Challenge for Change

Part One
THE GLOBAL DILEMMA

Chapter 1

Our Puzzling Stalemate

Recalling the gallant war effort, *Winston Churchill* coined the phrase, "Never before have so few done so much for so many!" Today, a mere four decades later, people all over the world ask themselves another question—"How did so few manage to mess up so much in such a short period?"

Evidently something has gone awfully wrong with our world. In spite of the chastening effect of World War II, Homo sapiens has demonstrated a pitiful inability to put his house on Earth in order. As he approaches the close of the twentieth century, man is facing the most critical stage in his evolutionary path. A new, portentous period of global confrontation has replaced earlier years of timid maneuvers toward detente. Star Wars and nuclear winter scenarios are our daily companions. Aching under the burden of a murderous foreign debt, Third World countries are worse off than they were ten years ago. Demographers and nutritionists paint gory future horizons because of the population explosion in most "have not" countries and insuffucent increases of food production. The rapid advance of microelectronics is threatening job opportunities in nations where chronic unemployment is already a dramatic scourge. Pollution is busily ruining our ecosphere, while economists despair over the queer ups and downs of the world economy and its risky teetering on the brink of bankruptcy, a behavior they are at a loss to explain or to predict. Political shortsightedness and rampaging technology seem to be pushing our civilization nearer and nearer to a holocaustal downfall.

By a strange coincidence, mankind is faced today by an impressive spectrum of different crises, all of which are thriving at the same time: The East-West crisis, the North-South crisis, the economic crisis, the energy crisis, the environmental crisis, the crisis of terrorism and the culture crisis, effecting big and small, industrialized and underdeveloped, countries alike. Their existence is omnipresent and is pounded on our consciousness with a brutal regularity that has transformed crisis and fear into a normal ingredient of our everyday life. Evidence of dismal disruptions and failures of national systems, and of the ex-

isting international order is flourishing everywhere. In many parts of
the world, the Angel of Death is reaping again a rich harvest. Wherever
we look, political tensions and social frustrations are building up, like
boobytraps, ready to go off and explode right in our faces. Little wonder
that prophecies of dangerous societal breakdowns and a cataclysmic
end of our civilization abound.

It is true the human race has survived agonizing trials before.
Even if the present aggregate crisis erupts in a deadly conflagration,
maybe a new and superior world order will arise from it like Phoenix
from the ashes. But never before has there been an overall crisis on
such a global scale. Never before was there an epoch marked by so
many virulent antagonisms, and with the nuclear time bomb ticking
away, threatening to blow up our planet at any moment, and with
hardly a streak of hope on the horizon that solutions to this messed-up
state of affairs may be in the offing. Somehow the world has stumbled
into treacherous moving sand, from which no escape seems to be in
sight. The global crisis, engulfing rich and poor countries alike, has
the features of a terribly entangled, hopeless stalemate, which only
tends to grow and to perpetuate itself like the coils of a vicious circle.
Arthur Koestler's recommendation to cure man's paranoiac tendency
via injections of a benign drug, is maybe the best illustration of the
grave illness of our civilization and of the staleness of some of our most
lucid minds, at the end of their wits.[1]

Doubts of the present and fear of the future are domineering traits.
Mankind is caught by the grip of a gigantic confrontation, in which
tendencies demanding a reshuffling of global structures and relationships are battling forces clinging to obsolete bastions and value systems. The present stalemate is but the product of this fundamental
conflict, whose final outcome remains a tantalizing incognito.

In the chapters that follow, I will dwell on some of the most glaring
manifestations of this stalemate. But it is not the global crisis itself
to which I wish to turn most of my attention. It is its intrinsic rationale,
the nasty coerciveness built into its operational system, that keeps the
stalemate stymied and shackled to its own dynamics like a prisoner
to his pole.

The crucial problem is not a shortage of profound analysis or diagnosis of the worldwide ills, but rather the dogged absence of politically feasible and consensus-capable alternatives. An amazing number
of highly specialized national and international institutions, and a host
of visionary minds, have applied over the past decades their talent to
the task of showing humanity a way out—to no avail. A devilishly
rational insanity keeps triumphing over the rule of restraint and reason.

I don't think that answers to this macabre dead-end situation will be forthcoming from wise and gracious extraterrestrials. These we will have to figure out for ourselves. But I do believe that contact with an advanced extraterrestrial civilization would have an enormously stimulating and beneficial effect on man's way of thinking and that, in the long run, it would profoundly influence our science, technology, and economy, and our global organizational order as well. The U.S. biologist *George Wald* once demanded that mankind should lick its problems without alien interference.[2] But once contact is achieved, two-way communications are bound to place at our disposal treasures of information that in due time, are bound to revolutionize and change dramatically the world as we know it today. Search of ETI and preparing man psychologically for one day meeting his peers, can help him bridge the dividing chasms, and bring him closer to the grand solutions that have eluded our practical capabilities and defied our bold imagination so far.

Chapter 2

Lost Faith in Politics and Politicians

As the past has ceased to cast its light upon the future, the mind of man wanders in obscurity.

—Alexis de Tocqueville

Let me begin by stressing a plight that highlights possibly better than any other our tragic impasse today. Though it has been left out of the UN's Declaration of Human Rights, I think people everywhere have a right to leaders and governments they can look up to with respect, pride, and identification. But how many politicians and statesmen are making people feel that way today? If politics is the art of furthering worthwhile human causes, it has been a poor art for some time. In rich and poor countries, and in liberal and authoritarian regimes alike, faith in the capacity of leading politicians to cope with the decision load, thrust upon them by the present-day crisis, has been waning steadily. This erosion of confidence in those at the top, in their wise guidance and statesmanship, their vision and abidance by a high moral code, has been going on for some time now.

It was not always like that. According to *Peter F. Drucker,* "a torrid political love affair (existed) between government and the generations that reached manhood between 1918 and 1960."[1] Then, in the wake of the first Cold War, the widening of the North-South gap, economic crisis, and the explosion of new reigns of terror and violence, disenchantment with politics began to smolder like a contagious plague. People's beliefs in the promises of national and world leaders, that they would usher in the golden age of progress, peace, and prosperity, a hope still very much alive during the early years after the big war, have gone down the drain together with the false prophets who nurtured them.

This deep-seated skepticism expresses itself in many ways. A typical symptom in the U.S. is the declining voter turnout and reservoir of party loyalty over the past twenty years. "In 1978," reports *Thomas W. Madron,* "the voter turnout for congressional elections was only 34 percent .... In 1976 (only) 54 percent cast a vote for President."[2] In

6

spite of *President Reagan's* landslide victory in 1984, the percentage of those eligible, who actually cast ballots, slipped further to 51.4 percent. Similar trends are in vogue in other countries. In Columbia, *President Turbay* was elected by an appalling 27 percent of eligible voters in 1978. People just don't seem to care any more.

The student revolts in the late sixties were another symptom of the growing estrangement between rulers and ruled. The subsequent boom of the cult, drug, and beat culture was likewise a signal of the missing rapport between the official discourse and record of governments and people's expectations. The Vietnam debacle and Watergate sent government prestige in the U.S. tumbling to an unprecedented low. The many peace, ecology, and feminist movements, and the crusades for human rights and other worthy causes, too, are all rooted in the same soil of distrust and discontent, with which broad segments of the population view the doings and undoings of those holding power. "They sprang from needs and perceptions," states *Raymond Williams,* "which the interest-based organizations had no room or time for, or which they had simply failed to notice."[3] They are vivid demonstrations of strong popular undercurrents against party platforms and government policies, no longer considered conforming to peoples' aspirations and their interpretation of national interests and priorities.

Unrest is also seething in the Soviet Union in cultural and scientific circles against the rusty party bureaucracy, the abuses of the "Gulag-State," and the outworn ideological myths, covering up the system's glaring failures and suspect goals. In Poland, the valiant challenge of "Solidarity" unmasked the phony claims of the Communist leadership to popular support. The exodus of more than a hundred thousand Cubans during a couple of weeks early in 1980 and the streams of refugees pouring out daily from countries like Vietnam and Cambodia provide telling testimony to the same malady. The Berlin Wall is, of course, the most indicting monument to the profound abyss widening in Communist nations between the regime and the people.

In Third World countries, opposition to self-appointed leaders and ruling elites has reached explosive levels. The popular struggles against *Amin* in Uganda, the Iranian *Shah,* the *Somoza* dynasty, and more recently the ouster of *Duvalier* in Haiti and of *Marcos* in the Phillipines, illustrate the potential of violence bottled up in these countries. In many, as in *Pinochet's* Chile, in South Africa, and in Lebanon, resistance to stale reactionary regimes and fruitless tug-of-wars has already led to chronic bloodshed, civil war or near civil war confrontation, exemplifying the dangerous nonconformity of vast majorities with the political leadership in this part of the world.

There are only a few exceptions to this general pattern. Before Iranscam, *President Reagan* enjoyed a high wave of popular support for revitalizing the American society and restoring part of the lost confidence in the leadership of the White House. In some underdeveloped countries, too, charismatic leaders such as *Castro, Gaddafi, Khomeini,* and *Ortega* hold part of their poor masses, always prone to embrace revolutionary gospels and fall for demagogic rhetorics, under their spell. But even the spell of the most charismatic leader wears off eventually, certainly upon his demise, when—as the rapid demystification of the great *Mao* showed—the curtain is drawn, and the ugly facets of ineptitude, misgovernment, corruption, and abuse are revealed.

People have good reasons for being fed up with the traditional political hocus pocus and bungling crisis-management. On the U.S. domestic scene, the occupants of the White House in the seventies were, to a large extent, responsible for the distress signs of the American economy, recession, the slump of growth and productivity rates, the loss of competitiveness of U.S. goods, and skyrocketing unemployment. Crime and narcotics rates went high, while suicide, mental illness and alcoholism statistics kept reaching new records every year. At the same time, the deterioration of the environment continued. The overall decline in cultural confidence, highlighted by the escapist rush into the nirvana of exotic cults, was evident. The vision of a greater humanization of the American society, whose social achievements would match its technological feats, remained a haunting hallucination. Successive Administrations failed patently to give the "American way of life" a new moral meaning, relative to the issues and apprehensions confronting man today.

U.S. leadership has been even less inspiring on the foreign scene. Though the *Reagan Administration* has been able to bolster the conviction of Americans that it will stem the Communist tide, the final outcome of the grueling rivalry with the Soviet Union is still far from settled. Third World countries confront the United States with a mixture of envy, contempt, and rejection. "The Vietnam experience," held *Robert L. Heilbroner,* "has undermined every aspect of American life."[4] It destroyed the credibility of the U.S. as a progressive, anti-colonial force. By actively supporting the overthrow of the *Allende Regime* in Chile, the late *Hans Morgenthau* charged, "We (the U.S.) have transformed ourselves into the most outstanding counterrevolutionary power of the status quo in this world." Yet the U.S. continued to support *Somoza's* blood dictatorship and the *Shah's* reactionary rule in Iran. The victory of the Sandinistas led to the first Communist stronghold in Central America. *Khomeini's* triumph entailed the humiliating and excruciating Teheran hostage drama, and precipitated a rude weakening of the West in the strategic Middle East, capped by the Soviet invasion of Afghanistan. The meek response of Washington to this challenge and its feeble demarches regarding the Soviet-inspired military takeover in Poland in 1980, highlighted the mediocrity of the U.S. foreign policy.

A look at the earlier postwar period reveals even bigger foreign policy blunders. *Jean Francois Revel* rightfully scathes the ridiculously mild U.S. reactions to the Berlin Blockade, the Berlin Wall, the Soviet penetration in the Middle and Far East, and the Communist takeover in half a dozen African nations. The ruinous handling of the Cuban situation and the military intervention in the Dominican Republic that boosted anti-Americanism throughout the world are other examples.

At Yalta first, and then in Potsdam in 1945, it was the astounding lack of political perspicacity of *President Roosevelt* and *President Truman* that was responsible for the handing away of the Eastern European countries to the Soviets as spoils of war, and that laid the groundwork for the Soviet Union's menacing predominance over Western Europe, based on a divided Germany, that has overshadowed world politics ever since. The scandalous way in which the fate of the Balkan countries was sealed in Moscow in October 1944 is a sad page of recent history. On a half sheet of paper, the degree of Russian influence in Rumania, Hungary, Yugoslavia, Bulgaria, and Greece was checked off in a matter of less than five minutes. Stricken by an assault of bad conscience, *Churchill* commented to *Stalin:* "Might it not be thought rather cynical, if it seemed we had disposed of these issues, so fateful of millions of people, in such an offhand manner?"[5] He proposed to burn the paper, but his scruples were brushed off by *Stalin.*

Though the U.S. emerged from World War II as the mightiest economic and military power on Earth, possessing the monopoly of the atomic bomb, its leaders carelessly threw away the chance to impose a genuine European peace settlement. But this was only the prelude to even worse bunglings to come. What are we to think, for instance, of the political expertise of five successive American administrations who let U.S. dominance inexplicably slip from their fingers; allowing the Soviet Union, in less than forty years, not only to pull even, but to surpass the military strength of the U.S. in many fields?

Other countries have not been much more fortunate. *Churchill,* for all his outstanding qualities, fought for the preservation of the British Empire but was forced to witness its liquidation. *De Gaulle's* fancy dreams of "grandeur" almost toppled France into anarchy and chaos. And it was *Konrad Adenauer's* dubious distinction, for torpedoing the only chance of German reunification in 1953, by rejecting the offer of *Stalin* and sanctifying thereby the painful ethnic vivisection of this nation, performed with the blessings of all Western leaders, an operation that more than thirty years later, is still a throbbing wound in the heart of Europe.

From a global point of view, the past forty years provide an impressive string of lost opportunities of the leaders of the two superpowers to come to terms, for addressing jointly the clamorous issues of the world. Instead conflict escalation, pursued recklessly by both sides, has pushed mankind to the nuclear stalemate and desperate North-South dilemma, which it now faces.

All this corroborates the conclusion that the recent fate of our civilization has been guided by an amazing degree of political amateurishness and disregard for man's basic needs and aspirations. Our politicians have learned nothing from fatal power politics and reactionary immobilism of the past. The real problem is not the throng of political racketeers, unscrupulous peddlers of influence, and eager but false prophets who reach the pedestals of power. The real malaise stems from the misjudgements and errors of honest and respected statesmen, the naive *Chamberlains,* easily deceived by *Hitler,* the popular *Roosevelts,* convinced that *Stalin* "was a nice fellow," and the crusading *Carters,* believing that preaching the human rights issue would force the Soviets to cancel out their plans for world domination.

But why has the art of politics descended so low? Why are there so few exceptional statesmen around, capable of reading the real meaning of our time and acting courageously according to it?

Men of science have tried to shed light on these puzzling questions—to little avail. *Harold Lasswell* and *Bertrand Russell* held that politicians are basically conservatives, geared to maintain the status quo of obsolete systems. According to *Erich Fromm,* "egotism causes politicians to evaluate their personal success higher than their social responsibility."[6] They prefer the catastrophe on the horizon to the sacrifice they would have to endure today to avoid it. *Yehezkel Dror,* on the other hand, argues that governments act crazily because ill-prepared and uninformed statesmen "react irrationally" to the unexpected events of life.

Arthur Koestler provided another interesting explanation. Quoting *Pascal's* adage that "man is neither an angel nor a devil, but when he tries to behave like an angel, he transforms himself into a devil," *Koestler* traces the ills of our age to the fact that our tottering political systems sweep too many demonlike fanatics and charismatic leaders with a messianic mission to power who, in the fashion of *Stalin, Hitler, Amin,* or *Pol Pot,* subject whole nations to their spell and evil designs.

People hoped that as a consequence of World War II the lure of the strong charismatic leader had worn off and that the pendulum had swung the other way, favoring the political fortunes of efficient managers of the complex state affairs and devout champions of peace and democracy. But sterile politicking, maladroit crisis-management, and perpetuation of the global stalemate have brought traditional politics in discredit. Little wonder that the call for the strong leader has already begun echoing in many Western nations.

Koestler's argument rings a familiar bell in many Third World countries, where colorful revolutionary zealots have managed to seize absolute power. Their self-professed goal is to deliver their people out of the bondage of poverty, reaction and imperialism. But in reality, what have the *Sekou Toures, Gaddafis, Khomeinis,* and *Ortegas* really brought to them except floundering economies, grim sacrifices for footing the bill of costly prestige projects and extravagant involvements on foreign fronts, loss of liberty, and morbid personality cults?

In the communist orbit, the *Stalins, Ulbrichts, Titos,* and *Maos* have given way to a generation of less despotic and enshrined leaders. Their hold on power depends more on collective leadership and on genuine achievements for the people than on controls via terror, as *Krushchev* found out in 1964. His days in the Kremlin were numbered when his missile buildup in Cuba was called a bluff by *Kennedy,* and when his ambitious plans to turn the Siberian "virgin lands" into a grain paradise suffered disastrous setbacks. But with the only possible exception of *Kadar* in Hungary, their place in history is open to doubt.

It has been argued that in totalitarian countries, in order to stay in power, it is vital for the leader to take a strong stand. Revolutionary hotheads everywhere excel in conflict escalation and substitution of compromise by confrontation. But the argument also applies to the democracies. *Premier Chamberlain's* political demise soon after his capitulation in Munich in 1938, sanctioning the dismemberment of Czechoslovakia, which set the stage for World War II; and *President Carter's* stunning election defeat in 1980, largely due to the meekly managed Iran hostage episode, are cases worth remembering. By sticking to SDI and standing up to Communist penetration in Afghanistan and Nicaragua, *President Reagan* has been able to whip up support for his policies, as has *Prime Minister Margaret Thatcher* for her radical, though costly, solution of the Falkland Island conflict, but whether the interests of world peace are served by it remains an open question. World politics is still played like a compulsive brinkmanship game, with stakes as high as a nations's existence and a system's survival. But when the dust settles, there are hardly ever any winners, only losers. The art of using the right doses of restraint or the right doses of strength at the right time is still an art poorly understood and practiced by the world's political leaders.

But how come so many inept political personalities and dilettantes, poorly suited to deal with the contemporaneous problem load, get swept into power? The existing systems seem to facilitate the ascent of those whose wanton ambitions are neither checked by their mediocre qualifications nor their moral restraints. In Western democratic societies, it is seldom the wisest, the one with an immaculate record, and the one most experienced in national and international affairs, who gets nominated and elected. It is rather those best versed in the reckless elbow technique for power and in party intrigue, most ready to come to terms with special interest groups, and to backpedal on promises made to the electorate.

Or is it just, as *ex-chancellor Helmut Schmidt* emphasized, "In democracy . . . he gets elected who knows best how to please the public."

Arnold Brecht charged that representative democracy has long since become a farce. We are being deceived two times, first assuming that we elect a candidate when we only elect a nominee, and second believing that the elected president or congressman will keep his promises and represent our interests. With the aid of mass media, ruling elites make sure that their candidates get nominated and elected. They effectively attract the public's distrust to competitors, advocating unorthodox ideas that may jeopardize the power elite's sanctuaries and short-term interests.

The pristine Communist leader, wholly consecrated to the advancement and well-being of his nation, is a similar myth. *Stalin's* bloodstained record as a henchman of millions of kulags and dissenters within the ranks of his own party should have dispelled this illusion a long time ago. So should the many other crimes, show-trials of innocents, purges, mass deportations, and calamitous economic experiments in a dozen other Communist countries. But in spite of the de-Stalinization trend in most, enough is known of widespread corruption and moral bankruptcy even within the ranks of high government officials, to bury forever the naive story of the Communist leader as a superior, unblemishing being.

Prerequisites for a successful political career are conformity with the party line, adulation of those in power and one's own superior, defense of the bureaucratic apparatus together with the shrewdness for eliminating competitors and inspiring trustworthiness. Constraints on criticism and spontaneous initiative and the privileges, reserved for the elitists' *nomenclatura,* are safeguards, making sure that no one steps out of line and that the ideological premises and organizational structures, holding the system together, are upheld, no matter how shaky and unsustainable they may be. To escalate to the highest party and government positions is a most difficult and hazardous task for even those with the most lucid, unprejudiced, and innovative mind.

If this analysis comes anywhere near truth, the conclusion is not encouraging. Under whichever regime man happens to live today, he seems to have little choice. He is doomed to the rule of politicians and statesmen, whose qualifications for guiding mankind's destiny at this critical stage deserve serious questioning. His participatory voice in their election is either muffled or declamatory und he has even less to say when they should be impeached or relieved from their duties. A change of this institutionalized pattern is not in sight. On the contrary, the sharpening of the global confrontation facilitates the consolidation of the entrenched power elites and of those appeasing the frightened electorates with appealing platitudes and offering them short-term panaceas. Outsiders have no chance.

In the Soviet Union, *Chairman Gorbachev's* reform attempts may yet fail because of the party hierarchy's and *nomenclatura's* stubborn resistance. Even *Teng Hsiao-ping's* bold liberalization program has, as of late, been forced to take a step backward and to tighten again the ossified party reins. On the other hand, as the role of the State, also in the Western world, as crisis-manager; promotor of economic, scientific, and technological development; military buildup and space projects; and as champion of social progress, tends to increase, the grip of the traditional political forces on the power mechanisms of the State is bound to get stronger. Reformist movements, such as the ecologists, will find it tougher to assert themselves.

Another factor that helps explain the intellectual and political drought distinguishing the top echelons of world leadership, is age. *President Reagan* already celebrated his seventy-fifth birthday. *Mao* only stepped down when he was already senile, and *Teng,* still holding on, is over eighty-five. In the Soviet Union, the long Brezhnev era was followed by the shortlived leadership of two ailing old men, *Andropov* and *Chernenko,* too weak for unleashing the overdue powerful impulses of reform. The suspicion is not farfetched that the comparatively young *Gorbachev* only got the nod for becoming the number one because in the mausoleum of the Politburo no other eligible living relics were available for succession. *Adenauer, De Gaulle, Tito,* and *Ulbricht* were also already old men when they finally stepped aside or were relieved of power by death, as was the *Generalissimo Franco.* It would be in the interest of all mankind if an early obligatory retirement age were introduced for politicians.

In the final analysis, we come to the unsavory conclusion that we have the politicians and leaders we deserve. *Churchill* once drew the following picture of the British leaders in the 1930s: "Delight in smooth-sounding platitudes, refusal to face unpleasant facts, desire for popularity and electoral success, irrespective of the vital interests of the State."[7] I wonder how many of the political leaders who participated in the shaping of history over the past four decades escape Churchill's mordant comment?

In his last book, the late *Herman Kahn* struck an optimistic note, estimating that the worst may yet be averted. But after considering the caliber of those in charge of world affairs, I wonder if this optimism is justified. It is neither bad luck nor by chance that we possess the leaders we have. They are as much a product of the contradictions and inadequacies of the existing political systems as contributors to them. We will be stuck with them as long as we remain stuck with the shortcomings of political democracy and party mechanics. Many important political decisions hailed by contemporaries as landmarks of great historic significance—when reevaluated with the benefit of hindsight thirty or forty years later—have revealed themselves as fateful blunders. It is therefore legitimate to surmise that it probably won't take long until many of today's most celebrated policy decisions will merit the same cruel judgement.

Summing it all up, the diagnosis couldn't be sharper in focus. Over the past forty years, none of the big issues of our time has received an adquate answer. On the contrary, the global crisis has only grown in volume and virulence. Star Wars and nuclear winter scenarios have become our daily nightmares. The underdeveloped world has swollen into a powder keg that may go off any time. Western Europe has slipped to a second- or third-rate status, to a mere object of the contest between the two superpowers, with little left but nostalgic memoirs of past greatness and fears of the future. Even the United States, in spite of its still ebullient society and regained vitality, has no larger dream than the preservation of its economic and technological lead and its utilitarian gospel. As to the essential questions, "how life is to be given an adequate meaning (and) how the quality of life and experience is to be assured," even such a knowledgeable and acute analyst as *George F. Kennan* acknowledges that neither the West nor the East have been doing too well.[8] Not much of a record for the world's leaders to be proud of. A historical balance sheet with so many distress signs that it will be a miracle if the holocaustal blow-up of our world can be averted.

The large majorities everywhere are forced to play a resigned spectator role and to serve as loyal victims to the destructive stalemate game, carried on by the world's leading statesmen with unequaled virtuosity. In order to overcome its stifling effects, imbue people with a new sense of participation in politics, and help launch effective assaults on the prevailing deadlocks, I think ideas of great intellectual appeal, moral power, and mobilizing and integrating force are necessary. Search of extraterrestrial intelligence and contact with it may be one of these ideas. A superior civilization would surely permit insights in incredibly perfected forms of leadership and social organization that may strain our boldest imagination, but yet prove enormously challenging subjects for examination. Pushing forward search and contact activities would help galvanize the human tide and unleash the social forces capable of forging the needed profound political change.

Chapter 3

The East-West Bottleneck

We must expect a conflict of hegemony between imperialist superpowers in today's historical phase.

—Carl F. von Weizsäcker

Communism is a better machine for world conquest than democracy.

—Jean-Francois Revel

Male deer and moose, locked in combat, often die rather than give ground. Unwilling or unable to disentangle their horns, they eventually perish from exhaustion. The contest between the U.S. and the Soviet Union, now dragging on for over four decades, is caught in a similar deadlock situation. Neither of the two superpowers is ready to make the necessary concession for reducing its further escalation. Neither can afford such steps, afraid that the adversary might take advantage. In a nutshell, this is the tragic dilemma of the East-West confrontation. We are condemned to live with it and possibly die with it because of its unrelenting rationale.

It was different forty years ago. The U.S. and the Soviet Union, having borne the brunt of the crusade against Hitler's Germany, were close allies. Victory crowds milled in Times Square and before the Kremlin, and American and Soviet soldiers who met vowed lasting friendship to each other. Nazi Germany, which had devastated most of Europe, threatening to impose a diabolic racist dictatorship on the old continent and maybe the whole world, had been defeated mainly by the joint efforts of Americans and Russians. The future seemed to shine bright for the bonds between the two nations. Wartime collaboration, fostered by the common deed for a worthy cause, nourished the conviction that the United States and the Soviet Union would live together in peace for the well-being of the entire world.

These naive hopes have long since turned into bitter irony. Forty years after the victory toasts of *Gen. Dwight Eisenhower, President Truman,* and *Stalin,* both countries face each other as inexorable enemies in a deadly bid for world supremacy that makes the attempt of Hitler's Germany to subdue the rest of Europe to the iron rule of its boots look like the antics of a charlatan. *Tocqueville's* prophesy 150 years ago, that in the wake of the decline of Europe two non-European superpowers would arise, the United States and Russia, has proven surprisingly correct. The Cold War period brutally unmasked the shallowness of the war alliance. The deceiving spirit of "detente" and "peaceful coexistence" during the sixties and early seventies failed to camouflage the deep antagonisms that divided the two nations and blocs, threatening more than once to tip the scales of world peace.

In 1986, world armament expenditures soared to the unprecedented high of $800 billion. After *President Carter's* vain exercises in placation, the frenetic military buildups in both countries since 1980 have been pushing the world closer and closer to the nuclear precipice. According to the ex-Chancellor *Helmut Schmidt,* the situation in the world after the Soviet occupation of Afghanistan resembled very much the tense situation before the outbreak of the first World War. In the stirring book, *The Third World War, Gen. Sir John Hackett* envisioned that it would break out in summer 1985. Though events proved him wrong, for years the nuclear time bomb has been ticking awfully close to detonation point. A peculiar doomsday psychosis has taken hold of people. *Ex-President Nixon* believes that "World War III has gone on now for a third of a century," and many other keen analysts doubt whether the global conflict, holding captive all of mankind, can ever be settled by peaceful means.[1]

This development was not and should not have been totally unexpected. Those with only an inkling of Marxism-Leninism might have known that the East-West honeymoon was bound to be over as soon as the common enemy was defeated. Nevertheless, the Berlin crisis, the ruthless Communist seizure of power in Czechoslovakia and Poland, *Mao's* takeover in China, and the North Korean aggression against South Korea triggered in the U.S. and in Western Europe shock waves of apprehension, ire, and repulsion. When North Korean troops crossed the 38th Parallel in summer 1950, a friend of mine panicked. He fled head over heels from San Francisco to Marin County, convinced that Soviet bombs would soon wreak havoc on American cities. He was not the only one. Millions of people, especially in Europe, were stampeded into similar schizophrenic reactions. Fear that the world would soon be plunged in a dreadful confrontation between the Western allies and the Soviet bloc began to take a terrorizing grip on people's imagination everywhere.

This fear was not gratuitous. Preventive war—as we now know—had at one point been a serious consideration of some of *President Truman's* closest advisors. The expansion of the Korean War into China, proposed by *General MacArthur,* may well have led to a global conflict. So could the Cuban missile crisis and the involvement of the U.S. in Vietnam. If outright war was averted, it was because neither side dared to push the fateful bottons. But short of hot global war, the war for global hegemony was on. Some of the highlights of this undeclared war were: the crushing of the popular uprising in Eastern Germany and of the Hungarian revolt by Russian tanks; the Soviet penetration in Africa; the Cuban-sponsored attempt to set Latin America aflame; the chilling termination of the "Prague Spring" in 1968 by joint Warsaw Pact forces, as well as the involvement of the U.S. in the Vietnam conflict, and its invasion of the Dominican Island. They made it clear that a global struggle had broken out for world supremacy, making short shrift of the earlier war alliance and that this struggle between the new "duo-pole" was the domineering political fact, overshadowing the rest of the world affairs for times to come. Other nations have since then gained political stature, among them China, Japan, and parts of Western Europe. But the antagonistic conflict between the U.S. and the Soviet Union remains the predominant power constellation till today, to which all international relations, including the North-South confrontation, are subjected.

A main feature of the postwar period is the rapid advance of Communism and the West's belated and often frenzied response to the danger that this advance poses to its survival. As early as 1957, Khrushchev had boasted that by 1970 the Soviet Union would overtake the United States in GNP and per head production. He startled U.S. television audiences with his assurance, "Your grandsons will live under socialism too." This bombastic announcement was not taken very seriously in the Western world, then enjoying an unprecedented economic boom. But chronic ulcers sapped the efficiency of the Soviet economy, and Khrushchev's successor, *Leonid Brezhnev,* quickly abandoned this fanfaronade. But the impressive Soviet military buildup, the Kremlin's efforts to break the U.S. monopoly of the A-bomb and development of the H-bomb, only one year after the first public test of an American thermonuclear device, the launch of Sputnik I in 1957, together with its expansionist drive all around the globe, removed the last illusions regarding the grand strategy of world communism. The past forty years thus bear witness to a spectacular increase of the might and prestige of the Soviet Union, principally at the expense of the United States. Following faithfully the gospel of *Marx* and *Lenin* that communism, as a historically superior social system, is bound to supersede capitalism, the Soviets were clearly embarked on a global, grand scale strategy with the aim of hastening the West's descent into the grave, and unifying the world under the banner of the "hammer and sickle." The Soviet and Cuban interventions and Communist take-overs in Angola, Ethiopia, Zaire, and South Yemen, and the Soviet-backed Vietnamese invasions of Cambodia and Thailand were evident thrusts in this direction. The military occupation of Afghanistan was one of the latest moves in accordance with this scheme as is Moscow's continued support of Nicaragua's Marxist government.

It should be acknowledged, however, that contrary to many ingenious beliefs in the peaceful intentions of its leaders, the Soviet Union never did renounce its mission to spread Communism. Soviet leaders never abjured to what, after all, is the final goal of Marxist-Leninism—world domination. The new Soviet Constitution of 1977 declares "consolidating the positions of world socialism and supporting the struggle of peoples for national liberation" as explicit objectives of its foreign policy.

But the deterioration of East-West relations cannot be blamed on the Soviet Union's designs alone. The deep-rooted political conservatism and inability of the U.S. and most of its Western allies to grasp the moral lessons of the revolutionary impetus unleashed in the Third World, helped opening the sluices to the Communist tide to sweep in with its appealing catch-phrases of independence, economic planning, and social justice. The State Department's irrestrictive support of reactionary monarchies and unsavory dictatorships—as long as they were opposed to Communism—contributed decisively to the success of Marxist and Soviet-aided takeovers and the incessant escalation of the global confrontation. As a consequence, many poor nations, disillusioned by the West's incomprehension, joined the bloc of non-aligned countries or, jumped outright on the Communist bandwagon. But this is far from all. The amateurish foreign policy of "brinkmanship" of *John Foster Dulles* that precipitated German rearmament, the ugly purge of the Arbenz Regime in Guatemala, the clumsy British-French intervention in the Suez crisis, the inflexible U.S. support of *Chiang Kai-shek* in Taiwan, and the ill-conceived and mismanaged interventions and military invasions in Cuba, the Dominican Republic, and Vietnam, added as much steam to the boiling Cold War period as Soviet aggressiveness. According to a study of the Brookings Institution, the U.S. alone has used military force as a political instrument in 215 incidents between 1945 and 1975.[2]

Next to ideological warfare, the development of always more deadly and sophisticated traditional and nuclear weaponry and its deployment in a vast orbit surrounding the Soviet Union was also a powerful propellant of the raging conflict. Even more impressive was the Soviets' contribution to the arms race. It led to brusque changes in the world military equilibrium. Specialists agree that by the end of the seventies, the Soviet Union reached military parity with the U.S. NATO's Dual-Track-Decision in 1979 to start installing by 1984 an array of Pershing II and cruise missiles on Western European soil, was a further step in escalating the confrontation. In response, the Soviets increased the number of their SS-20 missiles, targeted on Western Europe, illustrating the lurid automation ruling the arms race. The most far-reaching, but maybe also most questionable intensification of this unchecked race is, of course, *President Reagan's* Strategic Defense Initiative, better known as Star Wars, whose military and strategic implications have acted as decisive stumbling blocks at the recent *Reagan/Gorbachev* talks.

There have been attempts to check this dangerous and costly escalation of the conflict between the two superpowers. It was to be stopped by detente, by a policy of mutual restraint, based on the conviction that de-escalation and peaceful coexistence were the only rational alternatives to all-out war. The feeble agreements reached between *President Kennedy* and *Khrushchev* were first steps. *President Nixon's* summit talks in Peking and Moscow in 1972, which led to the normalization of Sino-American relations, the U.S. pullout of Vietnam, Salt I and Salt II, the reduction of Soviet troops in Eastern Europe, and the Helsinki Agreements in 1975, were already tangible results. In Germany, *Willy Brandt* launched the conciliatory "Ost-Politik," and even a Cuban-American rapprochement seemed not entirely unlikely. Token acceptance by the Soviet and American leadership to "refrain in their relationship from efforts to obtain unilateral advantages at the expense of the other," and of "linkage," in *Henry Kissinger's* terms, "an overall strategic and geopolitical view," based on national interests and the rule that no concessions should be granted without an adequate *quid pro quo* from the other side, had apparently made these advances possible. An encouraging reduction of the East-West conflict seemed to be in the offing.

Since then, however, detente sounds a tune of dead leaves. In the wake of Watergate, the Soviet takeover of Afghanistan, the aggressions of Vietnam, the recrudescence of the battle lines in the Near East and in Central America and the accelerated buildup of NATO and Warsaw Pact forces, detente has been swept into the trashcan of East-West relations. Gone are the days of eventual U.S. ratification of Salt II, of *Kissinger's* diplomatic acrobatics, negotiating the end of the Vietnam

War and compromises in the Arab-Israel crisis and of joint space projects, symbolizing Soviet and American scientific collaboration. Hopelessly stalemated reigns of war and terror give the political situation in the Middle East and in Central America their distinctive mark. The clash over Afghanistan is far from over. The resumed disarmament talks in Geneva have bogged down in the quagmire of mutual recriminations. After the failure of *President Reagan* and *Gorbachev* to reach a sensible agreement on Star Wars in Reykjavik, it is evident that irreconcilable differences continue to thwart the recent new feelers for de-escalation, highlighting once more the Gordian knots of the East-West bottleneck.

What are the chances for a reversal of this suicidal rivalry between the two superpowers? My appraisal leaves little room for optimism. The East-West confrontation has reached an appalling cul-de-sac situation.

Though I don't think that a hidden rationality is guiding history, we may ask the provocative question, whether this global contest is not a normal and maybe even wholesome process. *Konrad Lorenz* once rebuked the assertion that "war is the father of all things," but admitted that "with a better right this honorific might be given to conflict."[3] The ongoing East-West conflict may provide humanity with the necessary stimuli for overcoming the status quo and reaching a higher form of social organization. While this is probably true, the benefits which mankind is drawing from this conflict, holding the whole world hostage to the nuclear threat and to economic disaster, are conspicuously dubious. The prospect that an all-out nuclear exchange may settle the issue and that from its radioactive shambles—a hundred years later—a superior world order may arise, is not a very reconforting idea. But the indefinite dragging on of the East-West stalemate is hardly a more appealing alternative.

Unfortunately, most of the well-intentioned blueprints for overcoming this stalemate fail because of their political infeasibility. Who does not agree with the postulate that for the sake of survival a rational relationship of genuine peaceful coexistence and competition should replace the ominous rivalries marking our time? *George F. Kennan* insisted that the problem of Soviet-American relations should be addressed "with the best, in the way of sobriety, objective analysis, steadiness, thoughtfulness, and practicality of purpose, that our society can produce."[4] But as the German Philosopher and physicist *Carl F. von Weizsäcker* has stated, "The aperture of reason in a crisis is comparable to the aperture of an enchanted castle in a fairy tale." Precisely this kind of reason and rationale, capable of paving the way to a lasting de-escalation of the East-West confrontation and establishment of a Soviet-American relationship based on mutual trust and peaceful collaboration, is so difficult to summon up. Leading American statesmen are as much prisoners of their own blinkers and compulsions imposed by the interests of existing power structures as their Soviet counterparts. Neither in Geneva nor in Reykjavik were *President Reagan* and *Chairman Gorbachev* prepared to offer the far-reaching compromises and concessions necessary for a genuine reduction of mutual animosities and war potentials. Neither had the required go-ahead from their power elites.

Future possibilities of a "global arrangement" may be better gauged, taking a look at detente, its past performance and probable future worth. One fact immediately hits the eye: If detente was meant to ratify a Soviet-American relationship, in which both sides would refrain from actions tilting the precarious balance of power, and pursue a policy of reducing tensions and furthering peaceful coexistence, detente was at best a farce or, as *Jean F. Revel* holds, a "trap." The post-Kennedy years till 1978, generally associated with detente, witnessed a spectacular advance of Soviet power and influence. The Kremlin brought the rambunctious Czechs back into the fold and forced the U.S. out of Vietnam. It inflicted upon the U.S. a stinging defeat in the vital area of the Near East, during the Yom Kippur War in 1973, and gained new footholds in Africa and Latin America. It also managed to close rapidly the military gap with the U.S. For good reason, *Jean F. Revel* pokes mockery on the naive Western democracies, victims to the illusion that "they actually were in a period of detente," when "this was really the Soviets' cold war against them."[5]

Of course, detente helped the U.S. to press its own advantages. From accelerated German rearmament, strengthening NATO, helping to unseat hostile regimes as *Allende's* socialist experiment in Chile, and attempts to steer the smarting Middle East crisis to its own advantage, runs a long list of U.S. moves intended to counterbalance Soviet advances. The *Nixon-Mao* summit meeting in Peking early in 1972 which, according to *Kissinger,* ended the "bi-polarity of the post-war period and ushered in the present era of Sino-American collaboration, was evidently a bitter setback for the Soviets. It laid the groundwork to the almost airtight encirclement of the Soviet Union by hostile forces of the U.S., Western Europe, Japan and China."

That detente did not come near ironing out the underlying East-West antagonisms might have been clear from the beginning. According to *Kissinger,* "peaceful coexistence" was only "the most effective tactic" and "the best means (of the Soviets) to subvert the existing structure by means other than all-out war." *Khrushchev* himself admitted that "peaceful coexistence did not effect the continuation of the ideological struggle," and *Brezhnev* was even more outspoken, by declaring at the Twenty-fifth Party Congress that "detente does not in the slightest abolish . . . the laws of the class struggle," and that " . . . we see detente as the way to create more favorable conditions for peaceful socialist and communist construction." The West awoke only belatedly to the embarrassing fact that the Soviet interpretation of detente included not only the doctrine of "restricted sovereignty," applied vis à vis Czechoslovakia, but continued economic and military support to the anti-colonial struggle and the liberation movements in the Third World, as in Angola and Ethiopia. The Soviet occupation of Afghanistan, aid to the Marxist government in Grenada, and the continued support of the Sandinista Regime in Nicaragua underscore the strict continuance of the Kremlin's strategy. Relaxing Soviet-American tensions and easing the escalating East-West confrontation into reverse gear is therefore likely going to be an arduous and frustrating task.

First of all, powerful geopolitical interests pit the U.S. and the Soviet Union against each other. Geographically, the bulk of the Soviet Union is bottled up in one of the world's most unfavorable regions. It is encircled by an array of hostile forces, whose combined military and economic potential exceeds its own by far. In addition, it still suffers from an obsessive inferiority complex, rooted in near defeat in World War II and in the continued economic and technological superiority of the West, which helps explain the disproportionate anxiety of the Soviet leadership for its own security and defense. "The Kremlin," says *Robert F. Ellsworth,* a former Deputy Secretary of Defense, "seems to imagine a scenario of the worst case, a coordinated NATO-attack in Europe, a Chinese invasion from the East, and an extended rebellion in Eastern Europe, while the Red Army is still tied down in Afghanistan."[6] Though this may be a fantastic vision, it sheds light on the vulnerability of the Soviet Union's strategic situation. Since the Western powers and China seem determined to contain further Soviet expansionism and to maintain their strategic superiority, chances for defrosting the icy relations between Moscow and Washington are slim. A real breakthrough would require far-reaching reciprocal concessions, a withdrawal of Soviet troops from Afghanistan in exchange for a stop of Washington's aid to the Nicaraguan contras; or a simultaneous pullout of Soviet troops from Central Europe and of the U.S. from NATO, permitting the reunification of a neutral Germany. It is unlikely that the time is ripe for agreements of this magnitude.

A second roadblock to genuine de-escalation is the constant nuclear threat and the self-perpetuating nature of the armament race. Both superpowers threaten each other with devastating first and second strike capacities, capable of turning the greater part of theirs and other countries into charred and pestilent landscapes. Neither really wants the "unthinkable" happening, but neither can afford to trust the other side or let it gain a decisive advantage. Therefore, as *von Weizsäcker* holds, the rationale of each side is, and must be, to keep abreast of the other, militarily, strategically and scientifically. The deliriously spiraling arms race is the inevitable consequence of this absurd but in itself coherent rationale. Such is the automatized nature of this race, that nothing has been able to stop it so far. Since the political and ideological antagonisms between the two systems are likely to perdure, detente will probably perpetuate the basic conflict, carrying its rationale to ever higher levels of monstrosity.

Third, the sharpening of the ideological confrontation between the two opposing systems makes a Soviet-American rapprochement always more difficult. *Charles Bohlen,* an American expert on Russia, coined the apt phrase "the Soviet Union is a cause, not a country," and *ex-President Nixon* cut in the same notch, pointing out that, "The United States wants peace; the Soviet Union wants the world."[7] But the peace the U.S. wants to achieve is mainly the maintenance of unsustainable status quo situations, while the Soviets are committed to the goal of furthering the Communist cause. In this context, Marxist and Leninist ideology still plays a powerful role. It is mainly through ideological competition and penetration that Moscow hopes to gain superiority in the Third World, and to puncture the West's unguarded defenses. To believe that the Soviet leaders may be ready for ideological concessions is to hypothesize that they may be ready to cut the tree on which their might, prestige, and future is built. Nor may we suppose the West's willingness to accept a truce in the ideological warfare with communism. The intense ideological competition within the international communist movement reduces further the negotiable margin of the Soviets, making it necessary to tighten their doctrinarian orthodoxy and controls against penetration of bourgeois lifestyles and values.

A decisive battleground is the East-West competition for the souls and hearts of the large majorities in the Third World. Though in some countries the overall revolutionary impetus seems to be fizzling out, in a number of African and Latin American nations, Communist guerrillas are flexing their muscle, threatening to tilt the balance of power still further to the advantage of the Soviet Union. There is mounting concern that the weak slogans and halfhearted aid programs of Western countries may be losing out to the radical solutions proposed by Communist-inspired movements. *Thomas B. Larson* called attention to their appeal, centering "on programs for national development and escape from age-old backwardness and dependence on advanced countries."[8]

Such programs seem to have a greater impact than the Western catch phrases of freedom and U.S.-made consumer patterns. *President Reagan's* warm support of *Corazon Aquino's* ascent to power in the Phillipines, which hastened the departure of *Marcos,* and Washington's repudiation of the brutal slayings perpetrated by *General Pinochet's* security forces in 1986, may be signals that the times of irrestricted American support of reactionary dictatorship is over. But politically defined aid programs, capable of rousing broad popular support are still ostensibly missing. Under these circumstances, why should the Soviet Union want to abandon or reduce its support to the "national liberation struggle in Third World countries," a strategy that has given it supreme dividends!

As *Marvin Harris* has shown, in primitive societies two tribes will live in peace with each other, as long as they respect the integrity of each other's territory, wealth, and women, and do not permit demographic or ecological pressures to upset the equilibrium. In comparison, the East-West conflict is immensely more complex. Though women probably do not play a decisive role in this context today, the so-called *cordons sanitaire* and the zones of great strategic interest certainly do. The Soviet takeover in Afghanistan and *President Carter's* reaction to it, *General Jaruzelski's* coup in Poland, and the U.S. intervention in tiny Grenada, and fidgeting over the Central American caldron, illustrate the point. Neither the Soviet Union nor the U.S. are willing to tolerate hostile developments in countries vital to their security and defense. There are no territorial sanctuaries. Short of all-out war, a constant covert struggle continues to be waged with the intent to upset the existing balance of power. Even under detente, both sides have repeatedly reverted to this Machiavellian strategy and they will, no doubt, keep reverting to it in the future, though I must admit that the Soviets play this game more masterfully.

Basically the East-West conflict boils down to a radically different way of facing the future. While the U.S. is deeply pledged to maintain the capitalist order as it is, the Soviet Union is as determined to overthrow it and build a radically different world. Each is convinced that its system is the best suited for furthering the interests of man. Each is shackled to this obsession with all the compulsions arising from their political and economic structures, their ideological myths, and moral crutches, hostile to reform. *Andrei Sakharov* blames the hard line of Soviet foreign policy to the stern opposition of party apparatchiks and members of the *nomenclatura* to a meaningful liberalization of the Soviet system. In the United States, corporate interests and religious fundamentalism belong to the staunchest supporters of *President Reagan's* external belligerence. It is irrefutable that the greatest achievements of the Soviet state have been attained under Stalinism. And it was the need to face up to Hitler's Germany and Japan that put the prostrate U.S. economy back on its feet, just as *President Reagan's* gigantic defense programs injected new verve in the American economy, pole vaulting it to recovery from the severe slump into which it had lapsed in the early eighties. Both countries seem to be doing best when forced to measure up to powerful external challenges.

Convergence is therefore the least likely future scenario. The ongoing conflict is between two diametrically opposed systems and ideologies that admit no common ground or conciliatory terms. At stake is not only what each side stands for, but the survival of the system of each, its final worth before history. The fundamental rule is "winner takes all," a principle well understood by *Alexander the Great* and *Stalin*.

The conclusion from this analysis is not very optimistic. In default of an all-out nuclear showdown, the East-West confrontation is likely to remain the dominant contradiction of our world for a long time to come. Most likely it will just keep dragging on, until profound societal changes or upheavals in the U.S. and Soviet Union will render the confrontation superfluous or just make it fade away.

As of the beginning of 1987, a slim streak of rapprochement shone again on the horizon. The two meetings of *President Reagan* and *Chairman Gorbachev* in Geneva and Reykjavik made transparent that both sides consider de-escalation a desirable goal. The announcement that Soviet troops will soon be withdrawn from Afghanistan marks no doubt a major shift of Soviet foreign policy and represents a tangible political concession. But whether Washington will be willing to reciprocate, for instance, by scrapping U.S. aid to the Nicaraguan contras or compromising on the delicate SDI issue, cannot be taken for granted. It remains to be seen if *Gorbachev's* conciliatory gestures are more than just tactical ruses intended to anesthetize the West and to press overdue reforms at home and whether *Reagan's* acquiescence to his overtures will only be motivated by his desire to get Iranscam off his chest. So far, nothing warrants that the Soviet Union is ready for concessions regarding more substantial issues such as disarmament, reunification of Germany, and foment of world revolution, and nothing indicates that the U.S. is willing to adopt in its foreign policy a less schizophrenic attitude toward progressive regimes in the future. "It is vital," plead the authors of a recent study of the Aspen Institute, "that East and West alike resist the siren of superiority."[9] Whether this is a viable proposition remains questionable.

On the other hand, *Alexander Solzhenitsyn's* categorical assertion that "Communism will never be halted by negotiations or through the machinations of detente," fails to meet the issue.[10] The task facing mankind is not how to halt or destroy communism, which would mean war, but how to change it and give it a more human face. For this purpose, rapprochement, de-escalation, and detente are the only practical tools at our disposal. Lusty brandishments of sophisticated weaponry only make sense if their real objective is to bring the other side to the negotiating table. But for achieving a genuine breakthrough in the stymied Soviet-American relations, mankind should not be satisfied with cosmetic brush strokes at the fringes of the global crisis, but aim at higher goals, tackling the real issues of our time.

The East-West confrontation affects the life of every human being. Efforts to reduce and overcome it should stop being the monopoly of elites. The broad masses of people everywhere need to get involved in them. Final success may hinge more on their active participation than on the clumsy political machineries of the leading establishments.

This is no small undertaking. To bring it about, it will be necessary to mobilize political, philosophical, scientific, and cultural motivations, and to bring them into the fold of powerful currents in support of this cause. Man's often too complacent conscience needs an invigorating shock, one that may imbue his resistance to the self-perpetuating East-West bottleneck with more than an earnest wrath. The subtlest strands of man's conscious existence must be brought in vibration on behalf of a new world, united in harmony and peace.

Search and contact of extraterrestrial intelligence may play an important role, furthering this developmental process. In its early history, a superior civilization may well have passed through a similar stage, when its destiny was suspended on a tiny rope between apocalypse and tantalizing stalemate. It would be a fascinating subject for study. But even in its absence, the endeavor of man to cross the infinity of space in search of higher forms of intelligence and political wisdom will enrich his insights enormously, and enable him to spark the long overdue change.

Chapter 4

The Deepening North-South Gap

There is a real danger that in the year 2000 a large part of the world's population will still be living in poverty . . . . Mass starvation and the dangers of destruction may be growing steadily—if a major war has not already shaken the foundations of what we call world civilization.

—Willy Brandt

Thinking back on my university curriculum in political science forty years ago, a glaring shortcoming comes immediately to my mind. The venerable professors in Graz taught us just about all there was to know in political economy, from *Adam Smith's* liberal "laissez-faire" theory and *Lord Keynes's* theory of full employment to *Schumpeter's* ideas on technological innovation and *Marx's* arid lucubrations on the inevitable downfall of capitalism. But strangely enough, nothing was said in their lectures about what was soon to evolve into one of the most critical problem-issues of our time—the chasm dividing the world into "have" and "have not" nations.

Since then, this development gap has escalated into today's acrid North-South confrontation. While concern and frustration are growing in the industrialized world because hunger and poverty continue to be the lot of two-thirds of mankind, in developing countries, bitterness and social unrest are mounting. There is awareness of the grave dangers that threaten both wealthy and poor nations, if the abysmal gap is not bridged soon. But in spite of the considerable efforts during three development decades and prodigious aid from the North, it is widening rather than diminishing. Little wonder that next to the East-West conflict, the North-South gap is the most ominous and vexing facet of stalemate.

Its causes are deeply rooted in the unequal distribution of wealth, resources, and scientific expertise in the world. Of the globe's population of 4.5 billion in 1980, only 25 percent lived in the industrialized North. The other 75 percent, about 3.3 billion human beings, had their existence in the poor South. Yet, according to the World Development

Report 1984 of the World Bank, 79 percent of the world's total output was produced in developed nations, only 21 percent in developing countries. The latter's share in the production of manufactured goods is only 10 percent, that of energy consumption a bare 20 percent. Their participation in the world's R & D efforts is even more pathetic. As was revealed at the UN Conference on Science and Technology in Vienna in 1979, 97 percent of all R & D is carried out by the industrialized nations, a mere 3 percent by the underdeveloped countries.[1] Also, most patents, states the Report of the *North-South Brandt Commission,* "are the property of multinational corporations of the North, which (colaterally) conduct a large share of world investment and world trade in raw materials and manufactures."

Even more revealing is the social plight in the Third World. As was estimated by the World Bank, in 1980, some 800 million people lived in absolute poverty, with yearly incomes below $135, a standard judged by *Robert McNamara* "below any rational definition of human decency." These 800 million, almost a fifth of the world population, suffer constant malnutrition, live in precarious dwellings without sanitary facilities and with hardly any access to medical care and education. They belong to the bottom 30 percent of the world's poorest, with only 2 percent of global income, while at the other extreme the 10 percent of the world's richest people control 40 percent of total income. While average per capita income in Western industrialized nations reached $5,100 in 1977, about 1.2 billion people in the poor countries averaged $150, thirty-four times less than their luckier cousins in the North. An analogous situation characterizes food consumption patterns. According to a FAO report of 1983, "40 million people die yearly of hunger in Third World countries." Because of the prolonged 1984–85 draught and dire malnutrition, the death toll in Ethiopia alone exceeded one million. The World Health Organization estimates that around 15 million children died in 1982 from malnutrition and infectious diseases. Regarding infant mortality, UNICEF found that 97 percent of it occurs in underdeveloped countries. Those lucky to survive in Africa and Asia, may look forward to a life expectancy of 45 and 56 years respectively, a far cry from the average of 70 years and more in the developed world.

Unemployment and subemployment is another scourge of the Third World. In many African and Asian countries, unemployment reaches 15 percent, subemployment up to 40 percent of the active population. In 1980, in fourteen Latin American and Caribbean countries 42 percent of the economically active population was unemployed. Another issue is the tremendous educational backlog. In 1978, there were

800 million illiterates in the world, most of them living in the hunger belt of the Southern hemisphere. *Raul Allard,* the director of educational affairs of the OAS, revealed that in 1980 there were 43 million illiterates of fifteen years and older in Latin American countries, 20 percent of all people in this age group. The exodus of millions of jobless and untrained peasants toward the big cities, where they swell notorious slum colonies, is one of the inevitable consequences.

But this is not all. The most disturbing fact is that this profound and danger-laden gap is almost sure to increase during the decades ahead. Already in 1977, a U.N. study cast doubt on the chance that the modest goals set by the International Strategy for Development might be attained. Global studies as *"Interfutures"* of the O.E.C.D. in 1979, the *Brandt North-South Report* of 1980, and the *Global Report 2,000* of the U.S. government struck even more ominous cords. "In the year 2000," concluded the latter, "the number of the poor will have increased. By every measure of material welfare—per capita GNP and consumption of food, energy and minerals—the gap will widen."[2] According to an FAO study of 1980, more hungry mouths will have to be fed in the year 2000 than today, but there just won't be enough cereal reserves in the wealthy countries to compensate for the enormous deficits in the populous countries in the South. The *Brandt Report* also concludes "prospects of hunger and starvation (for hundreds of millions) are increasing."

The same gloomy prospect holds true for world unemployment. Extrapolations of the International Labor Organization show that more than one billion new jobs will have to be created until the year 2000 in the underdeveloped world.[3] Over the next two decades, India alone will have to provide job opportunities for 8 million people every year, Mexico for 1 million—goals neither of the two countries will be able to reach. The upshot is that the number of the unemployed is steadily rising and that millions of frustrated and desperate young people will keep joining the human reservoirs, in which the germs of violence and revolution thrive.

The income gap, too, is bound to widen. If present trends continue, warns the *Global Report 2000,* "a group of privileged nations; North America, Western Europe, Australia, and Japan will average a per capita GNP of $11,000 in the year 2000; by contrast in the LDCs, the average will be less than $600."[4] Incomes in the rich North will be forty-four times higher than in the poorest countries in the South.

Hopes that this comparable deterioration could be halted have been chilled by the aggravation of the world economic crisis during the early eighties. Stagflation, forty million unemployed in O.E.C.D. countries, plummeting raw material prices, high inflation and interest rates, and a choke-up of foreign trade and international financial flows helped the foreign debt of developing countries soar to over $800 billion by the end of 1983, making short shrift of all illusions about possible improvements of the dire trend.

During this period, prices of almost all basic exportable products from Third World countries suffered severe drops. As a consequence, Latin American imports fell by almost 29 percent in 1983, after having decreased by 20 percent in the previous year. The International Monetary Fund estimated that "real output of the non–oil-developing countries grew at a rate of 1.5 percent in 1983, about the same as in 1982, and much lower than the average rate of 5.5 percent achieved in 1967–80."[5] But considering the rise of population in most of these countries, the report concluded, "These recent output trends imply a serious reversal of earlier gains in per capita income."

The desperate situation of developing nations is mirrored by the dramatic evolution of their foreign debt. From 141 billion dollars in 1973, it had skyrocketed to 810 billion dollars by the end of 1983. Brazil, the country with the largest debt, owing some $81 billion, saw its ratio between servicing this debt and its exports jump from 17.4 percent in 1971 to 68.4 percent ten years later. Mexico, with a debt burden of $80 billion, would have had to destine in 1983 almost all its oil-export earnings to pay only the due interest. But critical debt problems affected not only big countries. A study of the U.S. Data Resources Inc., disclosed that "in 1983, two-thirds of the eighty-five analyzed developing countries, foreign currency reserves were inadequate for guaranteeing imports during the next three months."[6] This situation was further aggravated by the unprecedented rise of the prime rate for loans of U.S. banks by more than 72 percent from 1978 to 1981. For Latin America alone, the rise of this rate by only one point meant additional service payments of 2.25 million dollars in 1981.

Some developing countries have been able to avert the worst by renegotiating their foreign debt. Others have barely managed to postpone bankruptcy by tightening their belts and accepting the stringent adjustments imposed by the IMF, implying drastic reductions of state expenditures and imports, reduction of employment, and substantial rollbacks of living standards. As a consequence, "during 1980–83, per capita consumption declined by 2 to 10 percent in countries as diverse as Argentina, Brazil, Chile, Ivory Coast, and Yugoslavia . . . that the number of manufacturing jobs in greater São Paulo fell by 13 percent between mid-1980 and 1982, and that the frequency of bankruptcies in Argentina and Chile increased by several hundred percent until 1982."[7]

Though, since 1984, some Third World economies have been able to initiate modest steps toward recovery, experts agree that effects of the 1980–83 crisis will still be felt until the year 2000 and beyond. By 1990, a CECLA study warns, Latin America's foreign debt will top 450 billion dollars, and its service will require 100 percent of the region's total export earning during that year. As many as 35 million people in the working age will be unemployed, and the income gap will have increased. The gap is bound to keep growing.

Summing it all up, an inescapable conclusion emerges. Prospects that the poor masses of the South will soon be able to bridge the gulch that separates them from the well-fed, well-clothed, well-sheltered, well-paid, and well-trained majorities in the North, loom bleaker than ever before. In the year 2000, one-fifth of the human race, enjoying high levels of economic prosperity and material comfort, will be surrounded by four-fifths, facing hunger, unemployment, and little hope for betterment. Inevitably, new explosions of social turmoil are likely to knock on many nations' internal political stability, and subject world peace to unbearable strains.

David Horowitz once wrote: "Poverty, the scourge of man from time immemorial, has ceased to be the remorseless and inescapable product of mysterious and uncontrollable forces."[8] But to do away with poverty and underdevelopment has proven to be an immensely more complicated enterprise than experts thought. The roughly 350 billion dollars of official aid, pumped by Western industrialized nations into the Third World from 1970 to 1983, were just a trickle, unable to engender a profound change for the better. Two United Nations Development Strategies for the sixties and seventies, each launched with much expectation, proved unequal to the task. For good reason, disenchantment and angry criticism of man's incapacity to cope with this staggering problem is rapidly mounting in both halves of the world.

Since the UN General Assembly in 1974, Third World countries have pressed for the creation of a *New International Economic Order* (NIEO). They argue that the present order, imposed by the rich nations, gives the developed world the lion's share, reproducing via unequal terms of trade, unjust practices of technology transfer, and unfair monetary and financial operations the existing inequalities. It is hard to disagree with the basic logic underlying this assessment. In his book, *The Inequality of Nations, Robert W. Tucker* stated ten years ago, "The traditional international system must be judged a system of inequality par excellence." Consistent with this line of thought, many Third World leaders and experts, such as the Pakistani *Mahbub ul Haq*, have pleaded for "deliberate de-linking of the Third World from its past dependent relationships with the developed countries," and the replacement of the "pursuit of elusive present-day Western standards and per capita income levels."[9]

The response of the North to the needs of the underdeveloped South has been double-faced so far. Many statesmen, among them *President Kennedy*, the promoter of the "Alliance for Progress" and *Willy Brandt*, who, in 1985, received a UN award for his outstanding work on behalf of the developing countries, showed sympathy and deep insight in their plight, urging more Northern aid for reasons of self-interest, not only of charity. Organizations and agencies of the UN system have continuously pressed the wealthy North to assume greater sacrifices and to agree to reforms of the economic order. But in terms of real concessions, the leading industrialized nations stuck adamantly to their vantage positions. They have blocked the stabilization of raw material prices, turned down pleas for removing or lowering tariff and non-tariff barriers, especially of manufactured goods, and sidetracked requests to liberalize the stiff terms of loans. Efforts to establish more just codes of conduct for transnational corporations, and more egalitarian terms for technology transfer, remain till today futile paperwork. To boot, official aid of leading industrialized nations as the US, Great Britain, the German Federal Republic, and Japan is still a far cry from the 0.7 percent of the GNP target set by the United Nations more than a decade ago. The *Reagan Administration* even went as far as cutting repeatedly its official foreign aid.

It was no surprise under these conditions that the NIEO debate was quickly stifled and relegated to third-rank officials by most Western nations. The much heralded "North-South Dialogue" between top world leaders on the Mexican island Cancún late in 1981 only corroborated the fact that it had degenerated in a "dialogue between two deaf contenders." The lyric hopes of the *Brandt Commission* that the advanced North might reduce its astronomical armament expenditures and increase its financial aid were quickly washed down the drain, as issues as NATO's Double Track decision and Star Wars, and the refloating of the West's tumbling economy gained top priority. By the end of 1986, the North-South dialogue has slipped dangerously close to oblivion.

Little wonder that criticism of the North's self-centered policies has been growing, even in developing countries aligned in the Western camp. In the *Declaration of Quito* of January 1984 by twenty-three Latin American and Caribbean nations, the rigorous terms of the IMF for debt renegotiation and the policies of the North, forcing developing countries to bear the brunt of the world recession, were subjected to stinging criticism. Celebrating the thirtieth anniversary of the *Con-*

ference of Bandung, representatives of eighty Asian and African nations exhorted the industrialized world "to remove protectionist barriers and steepen up aid to the LDCs." ONUDI working papers and delegates to meetings in Vienna and Geneva in 1984 excoriated "the growing nationalism and bilateralism of the developed nations," the "retroceding trend of industrialization in Latin America," and the "resistance of industries in the North to reorient the creation of new productive capacities in the South." Regarding the stiff adjustment policies imposed by the IMF, *Germani Corea,* Secretary General of UNCTAD, charged in June 1984 that "they mean that the Third World must restrict its economies, while their poor majorities must lower still further their standards of living, in order to pay their foreign debt . . . ."[10] Responses of the North at the ONUDI meetings that it would be "utopian" trying to change the world order, underlined once more the reluctance of the industrialized world to acquiesce to substantial concessions, much less a crash program for the poor of the world.

But the meager results of the so-called "war against poverty" and the deterioration of the development gap cannot be blamed on the North alone. Ruling elites in developing countries also carry a heavy load of responsibility. *Richard A. Fagan* does not stand alone with his statement that most of these elites "bargain hard for fairer shares internationally, while doing little to increase equity at home."[11] It is mainly their fault, maintains the UN official *K.K.S. Dadzie,* if "the trickling down of modernization and progress from the rich to the poor countries has not reached the poor strata of these societies." The industrialized donor nations, which alone in 1983 transferred $33.6 billion in total aid, can hardly be blamed for the rapaciousness of Southern economic and military power groups. Though it is probably true, as *Robert Ramsay* from UNCTAD suggests, that "there is a strong community of interest between the rich people of the rich countries and the rich people of poor countries," and "as long as people are free to enrich themselves excessively, they will continue to apply excessive pressure to preserve a status quo that enables rich people to get richer," the burden of responsibility, carried by the wealthy of the South, is a heavy one.[12] Latin America provides many examples corroborating this harsh indictment. In many countries, the lukewarm response of local conservative elites to long overdue structural and social reform is the main culprit of the outbreak of violence and revolt that besets the region.

The "dollar flight," occurring especially in Latin American countries, where it is estimated that a third of its foreign debt could be paid with the funds that corrupt officials have deposited in private accounts of U.S. and other Western banks, is another story. It is estimated that during the last decade at least twenty-five billion dollars emigrated from Mexico to the U.S., and that from the eighty billion dollars of private loans to Latin America, some fifty billion returned to industrialized countries in the form of investments, acquisition of real estate, et cetera. The billions accumulated at the expense of the people by the former Philippine dictator *Marcos* and deposited and invested in the U.S. is another telling case.

The large expenditures destined by Third World countries for military purposes also sap their development capacities. In March 1982, the German Bundestag revealed the fact that social and economic development in the Third World was delayed and impeded because developing countries had spent more for armaments in 1980 than they had received in terms of official aid from the industrialized nations.[13] Aid totaled $35.5 billion, but armament purchases had amounted to $38 billion. The oft-repeated argument of Third World spokesmen, the industrialized nations should reduce their bulking defense budgets, therefore ought to be applied also to their own countries.

Aid, quipped *Robert Ramsay,* "is taxing money out of the pockets of the poor in the rich countries in order to fill the pockets of the rich in the poor countries." Another sarcastic bon mot may be "development is taking the money belonging or destined to the poor in the poor countries and handing it to the rich in the rich countries."

But widespread corruption, the dollar flight, and rearmament—and we should add squandering of limited resources for unproductive investments—are but some of the shortcomings the developing countries themselves have to account for. There are other paradoxical facets, for instance, the quasi-immunity enjoyed by the notorious "narco kings" in several South American countries until recently, which raised not a few suspicians, whether this was not one way in which the fabulous narco-market of the U.S., reaching $110 billion in 1984, was allowed to offset the negative balance of trade of these countries. Continued underdevelopment may not be laid exclusively to the doorsteps of the rich industrialized world. The truth is rather on the side of *Julius Nyerere,* the President of Tanzania, who courageously admitted in 1979, "If we don't work actively ourselves, eliminating absolute poverty among our own citizens, we have no right to complain about the gap of poverty and wealth between nations."

At the beginning of the eighties, a new development strategy was enacted in an effort of a new attack on the lacerating North-South chasm. It called for concerted action to satisfy the "basic human needs of the poor, especially in the LDCs, more emphasis on rural development, health and education, and a higher degree of collective self-reliance. But the vicious effects of the world's economic crisis, slumping rates of growth and trade, the strangling grip of the foreign debt evolution, protectionism and the dearth of financial resources confronted the implementation of this strategy, and the demand for more aid from the outset with insurmountable difficulties. The bill for oil imports of non-OPEC Third World countries rose from eight billion dollars in 1973 to over seventy-eight billion in 1980. The grain shortage of developing countries hit the appalling mark of eighty million tons in 1981, threatening to reach one hundred twenty million by 1990. The impossibility of feeding nine hundred million Africans by the end of this century has turned into a haunting nightmare of FAO officials. Major donor countries, hard-pressed by internal and other global priorities, kept applying a hard squeeze on foreign aid and throwing wrenches into trade liberalization.

Under these circumstances, which have only improved by the mid-eighties as far as the industrialized North is concerned, it is unlikely that the goals of the Third Development Strategy can be met. The target of reaching 25 percent of the world's manufacturing product by the year 2000, set for developing countries at the UNIDO Conference in Lima in 1975, is long since forgotten. Today UNIDO officials are happy if they reach the modest target of 15 percent. Another UN goal, to reach a 20 to 25 percent share in the world's R & D activities is also bound to remain unfulfilled, as is the UNIDO target for more than doubling the percentage of the population occupied in Third World countries in the industrial sector by 2000.

The financial resources needed to help these countries attain the mentioned goals are staggering. As the report of the *Brandt Commission* summed up, just to take care of the poor countries' most urgent food problems, annual transfers of $8.5 billion would be needed until the year 2000. An additional $30 billion would be required yearly during two decades to reach UNIDO's target of 25 percent of the industrial product. Another $6.5 billion annually would have to be transferred to cope with energy shortages and help with mineral developments.[14] If other urgent necessities as the development of R & D potentials, education, health, housing and protection of the environment are considered, the report estimates that the financial resources needed from the North would exceed one and possibly two trillion dollars only during the span till the year 2000. Evidently, pleas for transfers on such a scale are outcries into a stormy night to a deaf neighbor.

The conclusion we may draw from this analysis is that the North-South confrontation resembles a raft shooting down a gorge, headed straight for a descent the size of the Niagara Falls. Like the East-West conflict, it has so far defied all rational approaches and solutions. Leaders of the North and South readily agree that in the long run, the deepening cleavage between the two worlds can only bring disaster to both. But this is as far as the agreement goes. In November 1984, the Social democratic fraction of the German Bundestag proposed a new *Marshall Plan* "to overcome the North-South unbalance in the world economy." A "Fond for Development" was to be created, financed with substantial reductions of armament expenditures and bolstered by collaboration between the Western industrialized world and the socialist countries. The proposal was killed by deadly silence in all Western capitals. *Fidel Castro's* drastic proposal in June 1985, urging Latin American debtor countries "to go on strike," by refusing to pay their foreign debt, met—perhaps just as well—the same fate. A ghastly impasse-situation seems to have been reached. All recent negotiation rounds at the level of the U.N. ended in stalemate, in the midst of an irreconcilable contradiction between the South's poor performance and inflated exigencies and the North's ill-concealed parsimoniousness and adherence dictated by sheer self-interest.

It is only too natural that in this atmosphere of mutual distrust and recriminations the traditional pattern of development aid has been subjected to mounting criticism. *Prof. Melvyn Krauss* of New York University propounds the thesis "that stopping development aid would have a mobilizing effect." Even more outspoken are professors *Lord Bauer* and *B.S. Yamey* from the London School of Economics, who have indicted development aid as being "unreasonable, arbitrary and ineffective, even harmful to donors as well as recipients." Similar criticism has been vented in the Federal Republic of Germany. But the most scathing attack comes from the Indian "Birla Institute of Scientific Research," which in a study published in fall 1984 arrived at the astonishing conclusion that "the sixty-two billion dollars of foreign aid, which India channeled into its five Five Year Plans, have done more damage than benefits." The huge U.S. emergency aid in wheat, the study maintained, "had depressed prices of foodstuffs and had this way throttled the proper initiative of farmers to augment their production."[15] Findings as these shed light on the controversial state of the present North-South debate.

Though it was studded with some lyrics, it was generally agreed that the above mentioned report of the *Brandt Commission* contained some reasonable and useful proposals. Only the *Baker Plan,* by inducing a fall of the overvalued dollar and a decrease of interest rates, brought some temporary relief to the developing world. But prospects for the future loom bleak. The hostility that still breezes across the oceans and the basic antagonisms between the two superpowers makes a genuine defrosting of East-West relations, which might entail reductions of the armament burden and lead to increased aid to the developing countries, highly improbable. It won't spur the moral obligation of the advanced nations toward their poor relatives in the South. Third World countries will have to keep carrying the heavy cross of underdevelopment primarily by themselves.

This pessimistic appraisal poses two basic questions. First, what are the main causes of this perplexing dead-end situation? And second, is there any, however dim, hope that at least our children might someday live without the stigmatizing pain in their chest, called by *President Mitterand* "the West's heaviest burden on its conscience," namely that two species, those who have it all and those who have little or nothing, populate our planet? In order to place the answers to these questions in a more realistic frame, it is necessary to dispel a few misconceptions and romantic notions.

First, the so-called "engine of growth" or "locomotive effect" thesis, held in high esteem by the *Reagan Administration,* boiling in the nutshell down to the belief that high growth rates and prosperity in the developed countries will automatically lead to recovery and faster growth of the reeling Third World economies, is a fairy tale. Recent studies of the OEDC and CECLA show that growth in the North is but one positive factor for stimulating growth in developing countries and that many other factors, among them terms of trade, settlement of the foreign debt, correct use of aid, and national development policies play an even more decisive role.

Second, the assumption that the capitalist system, private initiative, and competition, free flow of capital, and technology and free trade are best suited for engendering dynamic development in Third World countries, may be an illusion. It is precisely due to the unequal rules of the game, imposed by this system, and to the lacking aggressiveness of the entrepreneurial class and resistance to healthy competition, that conditions of underdevelopment refuse to bow, and that forms of global planning for ensuring harmonious and undiscriminating growth as well as social progress are probably indispensable.

Third, the tendency, still rampant in many developing countries, to blame exclusively the North for all their misfortunes, is an untrue and dangerous oversimplification. Not only does it tend to hide and justify the unwillingness of power elites in many developing countries to consent to due structural transformations and social change, but it also gives free rein to sterile demonizations of many worthy contributions and advances of the industrialized nations, especially in the field of new technologies, thwarting thereby effective assimilation and innovation, and an active and wholesome insertion of their economies in the international economic order.

Fourth, it is probably wishful thinking to hinge hopes on substantial hikes of official foreign aid of the industrialized nations. Their resource allocations, linked principally to the exigencies of the ongoing East-West conflict, clearly take priority over the demands of more handouts for partly questionable efforts to overcome the developmental hiatus. This situation, which is unlikely to change in the near future, increases the pressure for indigenous development efforts.

Fifth, a profound contradiction exists between the strong nationalism, clamor for independence, and call for preservation of cultural identity in the Third World, and the marked tendency of the world, propelled by modern science and technology, to evolve toward the "global village," distinguished by a growing universalization of production and consumption patterns and cultural values. Because of the dizzy pace of the scientific and technological revolution in the North, and growing disenchantment of developing countries with the economic models and cultural styles of either of the two systems, American Capitalism and Soviet Communism, this rift is sure to escalate more in the future ahead.

Sixth, to help in overcoming the dreadful plight of the development gap, even a substantial increase of foreign aid would not be enough. To lick the inequity problem, wealthy nations might have to accept a different international division of labor, radical changes of their production and consumption patterns, and a drastic reduction of their standards of living. It seems extremely unlikely to me that such concessions could be wrung from the tough bargaining North, and that the implicit sacrifices would be accepted by the majority of people in the industrialized world.

The apparently unsolvable dilemma of the underdeveloped world has, of course many other facets and causes. *Simon Kuznets's* argument that as Third World countries push forward their industrialization programs—depending on imported capital and technology—inequity necessarily increases, is particularly thought-compelling. It suggests that because of rapid technology growth in the North, rich nations simply get richer faster than poor nations do. High rates of population growth still make it tougher for the latter to catch up.

What, then, are we to conclude? On one hand, we can probably dismiss *E.E. Schumacher's* aphoristic assertion "the problem children of the world . . . are the rich countries and not the poor." In view of what is known in terms of mismanagement and sluggishness of Southern elites to promote change, this thesis just doesn't do justice to truth. On the other hand, the charge that the response of the industrialized world still doesn't measure up to the dangers and challenges inherent in the North-South gap is not unjustified. Because of internal compulsions, Western statesmen cannot but refuse to gear their aid policies as philanthropists to the drums of Third World crusaders for equity. The ensuing predicament was summed up by *Robert McNamara* in these remarkable terms: "When the highly privileged are few and the desperately poor are many—and when the gap between them is worsening rather than improving—it is only a question of time, before a decisive choice must be made between the political cost of reform and the political cost of rebellion."[16] The tragedy is that this choice keeps being postponed.

All this interjects a rather pessimistic and almost fatalistic note in the ongoing North-South debate. It is a sobering thought that at least two postwar generations have been singularly inept in handling this basic problem-issue. It is even more disheartening that the future, stripped of all wishful daydreaming, looks hardly more encouraging. As long as the world's profound political and ideological divisions continue escalating toward showdown time, North-South equity is probably bound to remain a fancy utopia. Failure to have addressed this issue with determination today will lead to even sharper social unbalances and more dangerous focuses of global confrontation and violence tomorrow.

If this forecast is correct, my recommendation to further search and contact with extraterrestrial intelligence as a means for revitalizing man's dulled perceptions on the North-South stalemate is a valid proposal. Our lack of success should not make us immune to the success of others. Another more advanced civilization may have passed through similar stages of development, when its species was still divided in advanced and backward groups and refused to agree on the prerequisites for an egalitarian global order. After making contact, one of the worthy insights we might gain is how this civilization managed to overcome this stage. However different their path may have been, it would show us that the task, no matter how difficult, is not impossible.

Chapter 5

The Tragic Spiral of the Armament Race

*The specter of extinction hovers over our world and shapes our
lives with its invisible but terrible pressure.*
>   —Jonathan Schell

All my means are sane, my purpose and my goals mad.
>   —Captain Ahab
>   in Melville's
>   *Moby Dick*

In his book *The Year 2000,* the late *Herman Kahn* predicted that the
United States and the Soviet Union would normalize their relations
in the last quarter of this century and that a strengthening of world
peace would be the result. But trends in the late seventies and first
half of the eighties have not corroborated *Kahn's* optimism. With less
than fourteen years to go until the end of the second millenium, world
peace is a farce and probabilities of a nuclear clash have not diminished.
Hopes that with the aid of detente and balance of power, the ugly days
of the cold war would never return have given way to global concern
that the buttons which can plunge the world in a suicidal apocalypse
may be pushed any time.

Considering the high pitch of Soviet-American antagonisms, the
only really amazing thing is why all this eerie megatonnage hasn't
gone off yet. It almost has. At the time of the Berlin crisis, as was
revealed by a recently declassified U.S. War Plan, a preemptive strike
was to be launched under the code name *Trojan* against seventy Soviet
cities and industrial sites with atomic weapons. The chips were down
again during the Cuban missile crisis, and at the climax of the Six-
Day War between Egypt and Israel in 1967, the U.S. and the Soviet
Union had been on the verge of a shootout. After the NATO Double
Track Decision in 1979, and the increased deployment of Soviet SS-
20s, relations shifted even closer to showdown time. After the breakdown of the intermediate range nuclear forces (INF) arms limitation
talks in Geneva late in 1983, the *Bulletin of the Atomic Scientists*
moved its Bulletin Clock, the symbol of the threat of doomsday over

humanity, to four minutes to midnight. Now it stands at only three minutes before blow-up time.

Since World War II, some 150 wars and armed conflicts have caused more than 20 million deaths. Early in 1983, five conventional wars and thirty-five other internal conflicts were leaving bloody trails in three continents. Little wonder that *Richard Nixon* wrote, "World War III has gone on now for a third of a century." According to a secret Pentagon report that leaked out in March 1983, it may all start with the fall of *Ayatollah Khomeini* and a Soviet invasion of Iran, followed by an attack of Warsaw Pact forces in Central Europe, and a North Korean invasion of South Korea.

World War III did not start on July 4, 1985, as *Gen. Sir John Hackett,* a former high NATO commander, envisioned in a widely publicized book. But the big nuclear bang, hardly ruled out by the *Reagan-Gorbachev* summit talks, may yet arrive. As things stand, its threat is likely to accompany us as a nightmarish sword of Damocles deep into the twenty-first century.

Though the memories of the last global carnage are still very much alive, the ghost of war is again with us. The percentage of people believing that a nuclear holocaust will engulf the world is rising. War hysteria ran high in the U.S., especially after the Soviet takeover in Afghanistan. An opinion poll conducted by *Louis Harris* in 1982 showed that 63 percent of the interviewees thought "that there is a likelihood over the next forty years that the Soviet Union will attack the U.S.," and 69 percent believed "the Soviet leaders would not hesitate to use nuclear arms, if they were desperate enough."[1] More recent polls have corroborated these findings.

Human history is, of course, one of intermittent war. The First World War took ten million lives. World War II casualties exceeded forty million dead. Since then twenty million lives more have been sacrificed on the altar of a dozen cruel war gods. In the brief historical interlude from 1914 till today, mankind has legitimized genocide and hardened its heart and soul to it. For years, Soviet and American military strategists have been toying with war scenarios that would turn most of the U.S., Europe, and the Soviet Union into a lunarlike crater landscape, in which human casualties would be counted in hundreds of millions, and maybe even a billion of dead and injured. Devastating nuclear ICBM, SLBM, and bomber attacks can today be launched almost instantly by Soviet and American forces, raining death and destruction on hundreds of military and industrial targets and cities in both countries, with an automatized precision dwarfing everything in the history of human warfare. A man carrying a black briefcase with the codes for the strategic command centers follows the American president round the clock and a special jumbo-jet, equipped to serve him as a flying command post in case of a surprise attack, is kept at his doorstep day and night. False alarms have placed U.S. strategic forces on top alert several times, and Star Wars and nuclear winter have become stock in trade perspectives which man today lives with. For the first time in history, our species has acquired the lethal capacity to destroy the economic and ecological base on which civilization is built. "The extinction of Homo sapiens," concludes *Mark A. Harwell* in his grisly description of war in *Nuclear Winter,* "is a valid scientific question, no longer merely the hyperbole of science fiction novelists."[2]

The meaning of all this couldn't be more absurd and tragic. The world, together with its underground and space, has been transformed into a gigantic potential battleground. More than a billion people in the North are condemned to serve as hostages of the doctrine of deterrence based on Mutual Assured Destruction (MAD), while the rest of the world's population gapes in awe and disgust. If the folly of man deserves a monument, it has already been constructed—the gory nuclear machinery for mass slaughter and annihilation.

What would war fought with thermonuclear weapons mean to mankind? Much research has been done on this harrowing question, furnishing us a fairly clear picture about what Earth would be like after a ravaging nuclear blizzard. In *The Fate of the Earth, Jonathan Schell* has given us a stirring account of the sufferings, grief, and privations to which the horror-stricken survivors would be subjected. ABC's movie *The Day After* presented a moving portrayal of the aftermath of nuclear war by showing the hypothetical havoc and pain wrung upon the small town of Lawrence, Kansas. The overall cost would be mind-numbing. In terms of death, destruction, and damage to the environment, such a conflagration would exceed by far all the losses and sufferings man has inflicted upon himself in all of his history.

In fall 1980, "U.S. ballistic missiles carrying 7,274 independently targetable warheads, of which more than 5,000 were sea-based" were aiming at strategic targets in the Soviet Union. Pointed the other way at strategic targets in the U.S. were some 5,000 Soviet nuclear warheads, about 76 percent of them ICBMs.[3] Not included in this count were the thousands of tactical weapons, also nuclear powered, for shorter range usage. All together, the arsenal of deployed and ready-to-use nuclear weapons was estimated in 1980 at more than 40,000, with a destructive potential of 16,000 megatons* or sixteen billion tons of TNT, about four tons for every human being. Since then these arsenals for "overkill" have been increased, and it is now held that the combined nuclear forces comprise some 50,000 missiles, bombs and other devices, enough to turn most of the U.S. and Soviet Union into smoking shambles.

Regarding the casualties and damage caused by this destructive potential, the world possesses quite accurate information. "A single MT airburst has a fireball more than 1.5 miles in diameter," and as *Henry W. Kendall* from M.I.T. adverted, "It will flatten almost everything over about 50 square miles.[4] The intense heat from the fireball," he explained, "will set fires over an area approaching 100 square miles and cause second degree burns over 250 square miles. Some 600 to 1,000 square miles would receive (radioactive) fallout, lethal to unprotected persons, and over an area of about 2,000 square miles, there would be substantial risk of death or incapacitation. An additional 2,000 square miles would be contaminated beyond safe use."

In a similar study of the Office of Technology Assessment of 1980, the authors estimated that a 1-MT explosion on the surface in either Detroit or Leningrad would "leave a crater about 1,000 feet in diameter and 200 feet deep," and that within the immediate ring up to a distance of 1.7 miles from the center about 70,000 people would get killed in nonworking hours and that hardly any structures will remain standing."[5] Total casualties of the dead and injured up to a distance of 7.4 miles would be 250,000 and 430,000.

Weapons with a much higher yield would, of course, augment human casualties and overall destruction considerably. A twenty-five MT explosion over Detroit, at a burst altitude of 17,500 feet, the OTA study concluded, would level downtown within a circle with a radius of five miles. Skeletal buildings would be left standing only at its fringes. "There will be very few survivors," the study states cryptically. The 1.1 million survivors would have to bury more than 1.8 million dead, and attend almost 1.3 million injured. Likewise, as *J. Schell* points out, a twenty MT bomb, exploded on the Empire State Building in New York, would produce a fireball six miles in diameter, covering most of Manhattan and parts of Queens, Brooklyn, and New Jersey and killing everyone within it instantly. Supposing the radioactive fallout were carried to other populated areas, *Schell* estimates that "this one bomb would probably doom upward of 20 million people, or

---

*A megaton (MT) is equivalent to the energy released by the explosion of a million tons of TNT.

almost 10 percent of the population of the United States."[6]

In the mid-seventies, the U.S. Defense Department analyzed the effects of a Soviet "counterforce" attack, directed only at missile sites, forty-six bomber airfields and two submarine bases. It arrived at estimates ranging from three to sixteen million deaths. But this very conservative estimate had to be upgraded, because a preemptive Soviet strike would not limit itself on military targets but would also take aim at major industrial sites and main C^3I centers,* seldom located far from large urban areas. U.S. fatality estimates now range from a high of 155 to 165 million to a low of 20 to 55 million. These estimates do not include the number of injured and of deadly fatalities following the first thirty days after the attack. Vice versa, DoD studies assume Soviet fatalities ranging from a high of 64 to 100 million to a low of 50 to 80 million of the population. Better protection in the Soviet Union explains the difference. It is held that such a retaliatory U.S. attack would cripple the Soviet Union, destroying "70 to 80 percent of its economic worth." The authors doubt, on the other hand, whether after the Soviet first strike, "the U.S. would ever recover its position as an organized, industrial, and powerful country."[7]

Nuclear attacks with a megatonnage equivalent to several hundred thousands of Hiroshima bombs would therefore entail a massacre, making World War II look like a harmless skirmish. But even smaller scale nuclear exchanges would cause monstrous consequences. It has been estimated that with 1,200 forty kiloton warheads,† an attack of only nine Soviet submarines would bring death to 80 million Americans, and that an attack of twenty American submarines would produce a similar death toll in the Soviet Union. A Soviet strike, merely intended to wipe out America's missile force, is likely to kill some 15 million Americans. The survivors may derive little satisfaction from the fact that retaliating U.S. submarines would need only 1,350 Poseidon warheads to level just about everything to the ground in the 220 Soviet cities with populations exceeding 100,000, and that as *Henry W. Kendall* stated, as a consequence "the Soviet Union would no longer have national coherence," and that "its remnants could no longer function as a modern industrial society."

But death and destruction would not be limited only to the populations and territories of the main contending powers and their Northern allies. Many countries of the poor South, located in strategic areas, would also be targets of nuclear devastation. Others would suffer severe

*Command, Control, Communication and Intelligence
†Kiloton is the equivalent of a thousand tons of TNT.

losses due to radioactive fallout driven South by winds. Total human losses are consequently now assumed to be much higher than estimated earlier. According to a study of the World Health Organization of 1983, the total death toll from blast, immediate radiation effects, and firestorms in the targeted cities and military sites—including vulnerable non-Nato and Warsaw Pact areas—would reach up to 1.1 billion people.[8]

But not even these mind-boggling estimates reveal the magnitude of the catastrophe that would befall mankind in case of all-out nuclear war. It is now realized that the effects would not be limited to losses caused by blast, heat, radiation and fallout alone, but that long-term consequences on the ecological environment also had to be taken into account. Research findings, provided by outstanding American and German scientists, among them *Dr. Paul J. Crutzen* and *Dr. John W. Birks* as well as two scientific teams, headed by *Carl Sagan* and and the biologist *Dr. Paul Ehrlich,* unveiled the full scope of what is now accepted as the "Nuclear Winter Scenario."

In a nutshell, the Nuclear Winter Scenario boils down to the following. In the aftermath of a nuclear exchange in the range of 5,000 MT, gigantic quantities of smoke, dust, and soot would be hurled into the troposphere and stratosphere, obscuring the sun and producing a severe and prolonged drop in temperatures, which in turn, together with the long-term radioactive fallout, would entail disastrous consequences for agriculture, basic ecological systems, and biological processes. By assuming the "roughly a million square kilometers of cities would burn" within minutes of such an exchange, *Paul J. Cruitzen* estimated that about 200 to 400 million tons of smoke would be produced, with the result that "in an area between thirty and sixty degrees latitude in the Northern hemisphere . . . . hardly any light would be coming through." In the so-called TTAPS Study,* headed by *Carl Sagan,* it is estimated that these screens of black smoke would reduce sunlight to a small percent during weeks, and that it would take a year to regain normal illumination levels. Overall temperatures would plummet some 30°C, from $+10°C$ to the minimum of $-23°C$ and would only "return to the freezing point after about three months."[9] Similar findings were reported by the Soviet scientist, *Dr. Vladimir Aleksandrov,* who participated in the Conference on the World after Nuclear War, held in Washington in October 1983, in which leading Soviet and American scientists discussed the hazards of Nuclear Winter via a sensational satellite link-up. Temperatures, he stressed, would drop

*Derived from the initials of the authors' surnames—Richard P. Turco, Owen B. Toon, Thomas P. Ackerman, James B. Pollack, and Carl Sagan.

by 40°C in the Eastern United States and over Europe by as much as 50°C. "Eight months following the injection of dust and smoke into the atmosphere, the temperature in the U.S. and the Soviet Union would still be as far as 30°C below normal."

Regarding the effects of such prolonged periods of reduced sunlight and low temperature levels, *Paul R. Ehrlich* from Stanford declared, not only "would many animal populations be wiped out by the extreme cold," but "virtually all land plants in the Northern Hemisphere would be damaged or killed in a war that occurred just prior to or during the growing season. Most annual crops would (also) likely be killed outright."[10] The resulting 50 percent reduction of the ozone shield would increase exposure to ultraviolet light to dangerous levels for man and animal life and contribute to a drastic reduction of photosynthesis, threatening plant life with low productivity rates, degeneration if not extinction. Radiation levels, too, would cause lethal effects in a high percentage of exposed human beings, animals, and plant life in a large area.

These dramatic changes of the climate would terminate wheat production in Canada and most of Europe and move the corn production belt in the U.S. considerably to the South. In addition, it would seriously impair the chain of food production for the surviving people and animal stock.

Nor would the Southern Hemisphere be spared. Nuclear Winter effects are now estimated to produce profound ecological changes even in the regions of the tropical rain forests and to provoke serious disrupting effects on the precarious agricultural systems in many developing countries.

If all other aggravating variables are considered, ionizing radiation, toxic gases, inundations from burst dams, the spreading of epidemics, starvation, harmed water and sea ecosystems, the destruction of most transportation and communication links, the lack of adequate medical facilities, the breakdown of societal values and the return to instinctive survival patterns, a macabre conclusion emerges. Apart from the immediately dead and injured, exceeding a hundred million, because of secondary and long term effects, death would reap in the U.S. alone an additional harvest of 50 to 70 million dead. A similar death toll can be expected in the Soviet Union, most of Europe, and probably Japan. "On a worldwide basis," synthesizes *Mark A. Harwell*, "fully one billion deaths and one billion injured could ensue." And he concluded that as an upshot of a large nuclear confrontation, "modern civilization would at least temporarily cease to function in a recognizable way," and that "the fate of Homo sapiens would be literally left to the vagaries of nature reeling from unprecedented, anthropogenic insults."[11] The warning pronounced by *Donald Kennedy,* President of Stanford University, at the mentioned Washington Conference that "a major nuclear exchange will produce the greatest biological and physical disruption of this planet in its last sixty-five million years," and *Harwell's* admonition that it " . . . constitutes war waged on all peoples on Earth, war waged on the global environment itself," help grasp the magnitude of the unthinkable tragedy that would betide mankind.

But these findings of the effects of Nuclear Winter have not only aided to establish the proper proportions of the absurdity and monstrosity of the situation, to which more than four decades of the murderous spiral of the armament race have led up to. They also underscore the insanity of the continued East-West stalemate and that, in spite of precise knowledge of the unthinkable, the military behemoths of both superpowers and of many other nations continue to labor day and night, increasing the deadly potential and prolonging it indefinitely. In my doctoral thesis, forty years ago, I argued that nuclear weapons tended to render large-scale war obsolete. Now I realize that it was naive to think that the rationale which is moving world politics is rational by ordinary human standards. What the world is witnessing today is the wildest armament race that ever existed, a race which probably neither the Soviet Union nor the U.S. can win and that would mean Armageddon for both.

According to the Stockholm International Peace Research Institute (SIPRI), "world military spending in 1983" soared to "$600–650 billion," more than four times the amount spent in 1960.[12] The share of the U.S. alone rose from $60 billion in 1960 to over $186 billion in 1983, and the budget presented for 1986 hits the record mark of $318 billion. Soviet military expenditures are said to top these outlays by a considerable margin. Considering Star Wars and other new arms projects, the U.S. alone will spend about a billion dollars a day for defense and rearmament as early as 1987. Other nations are not lagging behind.

Third World military spendings, too, have been rising steadily. According to SIPRI, they grew by an average of 7–8 percent a year from 1970 to 1979, faster than their GNP. Their share in world military expenditure increased from 6.3 percent in 1965 to 17.7 percent in 1983 and in real terms from $66 billion to over $113 billion in 1983 (not including China), which exceeded by far the amount received by developing countries in terms of aid by industrialized nations in the same year. "About 65 percent of the total arms flow during 1979–83," revealed the same source, "consists of imports by the Third World." The lion's share of this lucrative international arms trade in the 1970s—called the Decade of Disarmament—belonged, of course, to the U.S. and Soviet Union, each with a share of about one-third.

Since World War II, the headlong arms race has cost mankind more than 10,000 billion dollars. Armed forces around the world total more than 25 million, and according to a UN study led by *Inga Thorsson,* over fifty million people are engaged in military related occupations, among them an estimated 500,000 highly qualified scientists and technicians, of which roughly nine out of ten are Americans or Russians. Over two-thirds of the total U.S. federal R & D expenditures are earmarked in 1987 for defense, totaling $43.3 billion. Comparable R & D efforts are carried out by the Soviet Union.

These shocking figures which, above all, provoke continuously the wrath of the poor two-thirds of mankind, show how futile efforts have been so far to bring the gigantic military buildup of the two super-powers under control. Neither the Vladivostok Agreement of 1974, whose aim it was to limit Soviet and U.S. strategic delivery vehicles to 2,400 for each side, nor SALT II of 1979, which attempted a further reduction of the number and range of offensive missiles—but was not ratified by the U.S. Congress—, nor the 1972 bilateral ABM Treaty, which banned the development and deployment of land-based Anti-Ballistic Missile systems, succeeded in curbing the rampageous arms race. In spite of all pious declarations to the contrary, it has continued during the eighties at a dizzier pace than ever.

Over the past few decades, the Soviet Union has succeeded in building up a formidable war machine. By the end of 1983, its strategic nuclear weapon capability was held superior to that of the U.S. in number of delivery vehicles and megatonnage. Its land-based SS-19 Mod 3 and SS-18 Mod 4 intercontinental missiles with a range of 10,000 and 11,000 kilometers and six and ten warheads respectively, could reach and destroy all major U.S. military sites and cities. The Soviets have now the world's largest ballistic missile submarine force, twice as large as that of the U.S., and possess an effective strategic air force, including 235 Backfire bombers, capable of carrying out missions to the U.S. mainland. The Soviet naval forces, composed of nuclear-powered Thyphon submarines and aircraft carriers, and modern guided missile destroyers and cruisers, make up a fleet, second in numbers of ships and worldwide mobility only to that of the United States. These strategic forces are complemented by a vast array of conventional and highly sophisticated smart weapons for tactical defensive and offensive use. In Europe, Warsaw Pact tank forces outweigh their NATO counterpart by a factor of 2.5 to 1 and powerful SS-20 missiles with a range of 5,000 kilometers, a CEP of 400 meters, and three 150-kiloton warheads each, bear down on targets from Narwik to Gibraltar.

The unprecedented expansion and modernization of the Soviet armed forces led to the equally impressive Western response. Spurred by the Soviet invasion in Afghanistan, the military takover in Poland, and facing a growing vulnerability of the U.S. to Soviet ICBMs and LSBMs and of NATO defenses, the *Carter* and *Reagan Administrations* began to match the Soviet military buildup. The central goals for regaining military and strategic superiority for the U.S. call for the new intercontinental MX missile, slated to substitute the old Minuteman IIs and IIIs; the construction of the B-1 bomber, programmed to replace the outworn B-52s; and the introduction of the Trident II D-5 submarine-launched ballistic missile, in addition to the Trident I missile system. This basic triad will be complemented by the deployment of new nonstrategic nuclear forces such as the Pershing II missile and the cruise missile; the XM-tank; highly sophisticated aircraft, such as the invisible "stealth" plane; a broad assortment of smart weapons, and upgraded C^3I systems.

Plans call for the construction of 100 MX-missiles with ten MIRV (Multiple Independent Reentry Vehicle) warheads each, at a cost of about $40 billion, and to be deployed in present Minuteman silos in the late eighties and early nineties. The B-1 bomber, scheduled to be introduced in 1986, will carry long-range cruise missiles, short-range ballistic missiles and, among others, the new 1.1 megaton B-83 bomb, which the pilot will be able to release at supersonic speed from as low as fifty meters off the ground. The U.S. Navy plans to go ahead with the construction of twenty to twenty-five Trident II submarines at a cost of $70 to 85 billion, each carrying twenty missiles with eight 475-kiloton warheads, with a range of up to 7,500 kilometers and the needed accuracy to crack Soviet ICMB silos. Simultaneously, some 140 units, from submarines to aircraft carriers are programmed to be equipped until 1990 with new Tomahawk sea-launched cruise missiles (SLCMs), a very flexible nuclear weapon with a range of about 3,000 kilometers, and distinctly offensive capabilities. For tactical warfare on the ground, a whole array of "smart" weapons, using "precision-guided munition," have been introduced or are in late stages of development.

On the European theater, the U.S. has been pushing the NATO Double Track decision of 1979, which called for deployment of 464 ground-launched cruise missiles and 108 Pershing IIs. By the end of 1985, the 108 Pershing IIs were deployed in the Federal Republic of Germany, and 128 cruise missiles in Great Britain, Belgium and Italy, from where they could hit Soviet targets, including Moscow, with a warning time of less than ten minutes.

Secondary powers have not stayed behind in this inevitable race toward always more deadly, sophisticated, and costly weapons systems. France has built up a "force de frappe," composed of bombers, nuclear submarines, and missiles, capable of launching already in 1982 more than 100 nuclear warheads with a combined yield corresponding to 4,160 Hiroshima bombs. Great Britain, too, has been busily rearming, modernizing its fleet, and reequipping its submarines with Trident II missiles. "By 1990, French and British submarines alone . . . will have a total of 1,268 warheads targeted on the Soviet Union."[13] The Federal Republic of Germany has developed the much coveted "Tornado"—at over $50 million a piece—and is engaged in development of such "smart" weaponry as surface-to-air and air-to-surface missiles. In 1980, China's nuclear arsenal included fifty to seventy IRBMs, forty to fifty MNRMs, and ninety Tu medium bombers and it has grown since. Japan, too, has as of late augmented its military buildup under U.S. pressure.

But this is not all. Since 1983, the momentum of the arms race has been shifted into a still higher gear by *President Reagan's* Strategic Defense Initiative, or Star Wars. Its objective is to build with the aid of laser-beam-equipped space stations a near foolproof umbrella against a possible Soviet nuclear attack. Powerful lasers and other death rays are supposed to destroy the oncoming missiles in a three-layered defense: first, during the booststage of the rockets, or soon after blast-off; second, high over the stratosphere; and third—if they survived the first two phases—during the final approach, as they home in on their target. The Star Wars program, presented to Congress for the five-year effort until 1990, calls for $26 billion, of which about $5 billion have been allocated for 1985 and 1986. Informal sources, however, place total cost till the year 2000 at anywhere between $100 billion and $2 trillion.

Whether the enormously sophisticated Star Wars project is technically feasible still remains a highly controversial issue. Secretary of Defense *Caspar Weinberger* exulted in December 1985 that such a laser system umbrella might be ready by the mid-nineties, earlier than had been expected. The successful test at the Kwajalein Atoll in the Pacific on June 10, 1984, where a missile equipped with an infrared sensor intercepted and destroyed a dummy Minuteman warhead, posing as a Soviet attacker, approaching at 25,000 kilometers an hour at an altitude of about 160 kilometers, seems to have confirmed the missile-killing capacity of the U.S.

That Star Wars might ultimately lead to a still more dreadful escalation of the armament race, as many politicians and scientists charge, is not an entirely gratuitous proposition. The late *Yuri An-*

dropov saw in SDI a threat "to disarm the Soviet Union," and it forced the Soviets to undertake feverish R & D efforts to counteract the project and to thwart the menace of an impenetrable U.S. ABM system, while remaining vulnerable to an American missile attack. If Star Wars will induce them to make substantial concessions in the renewed Geneva disarmament talks still remains to be seen. The Reykjavik summit meeting went to naught when *President Reagan* refused to budge on the critical SDI issue. The right of a country such as the U.S. to build up its defenses against the genocidal capacity of a potential aggressor is certainly legitimate. *President Reagan* even proposed that his successor could hand the technology of the successful ABM system to the Soviet Union, just to prove "that there is no point in both sides keeping bulging warehouses of these deadly weapons any longer."[14] But the Soviets were quick in rejecting this rather offbeat offer.

No less significant has been the parallel erosion of traditional deterrence as the basic U.S. defense strategy, and its replacement with the buildup of a "war-fighting capacity," a strategy—as many hold—meant to give the U.S. "a first strike capability." The departure from dependence on deterrence goes back to *President Carter's* Presidential Directive 59, signed in July 1980. Alarmed by the growing force of Soviet intercontinental SS-18s and SS-19s with gigantic twenty and ten megaton warheads and rapidly improving accuracy, Pentagon specialists were becoming convinced that the Soviets were building "war-fighting weapons," and that by 1981 and 1982 the latest, they might achieve a "kill-ratio" of 80 to 90 percent of Minuteman silos. The rationale behind PD 59, bolstered by the decision to build the MX hard-target killer missile and an effective C^3I system, was, as Defense Secretary *Harold Brown* put it, to convince the Soviets "that . . . no use of nuclear weapons—on any scale of attack and at any stage of conflict—would lead to victory." The *Reagan Administration* went a large step further. According to its First Complete Defense Guide, the new nuclear strategy calls on American Forces to be able to "render ineffective the total Soviet (and Soviet-allied) military and political power structure."It further postulates that in a nuclear war the United States "must prevail and be able to force the Soviet Union to seek earliest termination of hostilities on terms favorable to the United States."[15] The clincher of this program, incorporated in what is now called the Defense Department's Single Integrated Operational Plan (SIOP 6), is the assumption that the buildup of "a war-fighting capability of the U.S." will deter a Soviet nuclear attack. It is based on the development of the nuclear triad: the MX-missile, Trident II, and B-1 bomber. But it is also shored up by a $31.5 billion C^3I program and the establishment of a "strategic reserve force" by 1984, consisting of covertly deployed submarine-launched ballistic and cruse missiles and a widely dispersed bomber force. Part of it is also the introduction of a highly effective anti-satellite system by the U.S. Air Force by 1985, using "miniature homing vehicles," and, of course, the development of a vast array of laser weapons for Star Wars.

Top U.S. policy makers have denied the accusation that the purpose of this mind-boggling rearmament program may be the intention to bring the Soviets down to their knees with a devastating preemptive attack. *President Reagan* told the American Legion in 1983 that " . . . we have never sought nor will ever develop a strategic first-strike capability," and Defense Secretary *Caspar Weinberger* has stated the Administration's belief that "neither side could win a nuclear war." But the U.S. Air Force's *General Bernard Randolph* acknowledged to Congress already in 1982, "We want to be able to wage wars with the space systems," and government advisor *Colin S. Gray* reportedly gave vent to his conviction that "a decisive victory in space is possible."[16] Similar ideas have been espoused in Soviet military writings and probably dictate Soviet military schemes.

This new strategy evidently facilitates the escalation of a major conflict into a nuclear shootout. The danger is not that madmen in the White House or the Kremlin may purposely conjure up a showdown. It is rather, as *William M. Arkin,* director of an Arms Race Research Project of the Institute of Policy Studies in Washington has pointed out, that "the accumulation of weapons capabilities, together with improved planning, command and communications . . . are creating forces which will be perceived by their controllers as too tempting to let alone during a conflict."[17]

These new war-waging strategies make it painfully clear how frantically the two superpowers have been moving away from "cold peace" to the verge of "hot war." Furthermore, if it is born in mind that, as *Michael D. Wallace* has shown, out of a total of twenty-eight past arms races, twenty-three have led to war, few people will take exception with *George B. Kistiakowsky's* gloomy prediction in 1981 " . . . that the use of nuclear weapons in a local war is highly probable by the year 2000, and (that) once a nuclear weapon is used, a global holocaust is likely."

Disarmament has proven a dismal failure. Treaties since 1945, such as the "Partial Test Ban Treaty," "The Outer Space Treaty," "The Non-Proliferation Treaty," "The Biological Weapons Convention," and SALT II, among others, have barely scratched the surface of limiting the arms race and have failed conspicuously to roll back the threat of war. Underground nuclear tests for military purposes have continued and the Non-Proliferation Treaty has been violated with impunity time and again. Biological weapons have been prohibited, but the potentially more harmful chemical weapons have not. Both have allegedly been used in Vietnam, the still raging Iran-Iraq War, and in Afghanistan. The Intermediate Range Nuclear Forces (INF) arms limitation talks, that dragged on in Geneva for two years, finally stumbled over the questionable "zero option" (no intermediate missiles on either side), and because of reluctance of the U.S. to include in the count some 1,200 British and French nuclear warheads. Other negotiations, such as those on Mutual and Balanced Force Reductions (MBFR), regarding Nato and Warsaw Pact forces, that began in 1973 and the World Disarmament Campaign, launched in 1980 by the United Nations, also ended without tangible results.

Recent attempts to overcome this stalemate situation have been equally discouraging. The proposal of the Soviet Union of 1983 against "Militarization of Space" was filed away in the UN. The six-month moratorium on the deployment of new Euro-missiles, announced by *Gorbachev* in 1985, and prolonged later, was not reciprocated by Washington. Neither has NATO backpedaled from its Dual Track decision. Plans to transform the Balkan, the Scandinavian countries, and the South Pacific into "nuclear-free" zones have run aground, and the project, sponsored by German Social Democrats, to create in Central Europe a zone free of chemical weapons, was killed even before starting a controversy. The bilateral arms control negotiations, resumed in Geneva early in 1985, still stand—after several rounds—where they began.

Voices of other nations and of scientific groups against the rampant arms race were no more than lone cries in a dark night. The joint declaration of the *Brandt Commission* and *Palme Commission* in January 1984, asking governments to allocate part of their military expenditures for development aid and assuring world peace, was stoically ignored in Moscow and Washington. The New Delhi Anti-War Declaration, signed by six presidents and prime ministers in January 1985, pleading for the prevention of the arms race in space and a comprehensive test ban treaty, but failing—curiously enough—to speak out against conventional war, met the same ignominious fate. The correspondence of *Samantha Smith,* the young schoolgirl from Maine who died in a plane crash in 1984, with *Yuri Andropov* in 1983, and her innocent question "Why do you want to conquer the whole world, or at least our country?," probably stirred world opinion deeper than these stale declamatory gymnastics.

All this indicates that stopping the mad arms race and nuclear confrontation is a much more complex task than many apostles of peace round the world believe. Well-intentioned appeals against the immorality of nuclear incineration fail to impress military strategists. Third World reminders that for the price of one fighter plane, 40,000 poor villages may be supplied with modest medical supply; that for the value of one modern tank, a thousand classrooms for 30,000 pupils could be built; and that with the world's military expenditure of half a day only, the World Health Organization's program to eradicate malaria could be financed, have been no less futile moralizations. *Pierre E. Trudeau,* former prime minister of Canada, touched the crux of the issue, rapping the paradox that "politicians, who once stated that war was too important to be left to the generals, now act as though peace were too complex to be left to themselves."[18] The bitter truth is, in a world overwrought by crisis, distrust, hatreds and rivalries, disarmament must necessarily remain a stranger. As long as the East-West confrontation for world supremacy does not give way to genuine collaboration, the collision course must remain mankind's mainstream. Though it is now agreed that an all-out nuclear showdown would not mean the end of our species, it is clear that it would inflict staggering losses on mankind and that our civilization may be thrown back by as much as a hundred years. But even these apocalyptical specters have so far not brought our world to its senses. During the early cold war years, *Sir Bertrand Russell* could afford taking the large view, pointing out that a surgical definition of the confrontation one way or another was to be preferred to the indefinite dragging on of a nuclear stalemate. But today only genocidal desperation, bordering on insanity, can father the same proposal.

Four decades of nuclear escalation, however, have illustrated to the hilt that *Sir Bertrand Russell* had foreseen the big dilemma facing mankind ever since. Our civilization's most tragic weakness is its inability to find a way out from the deadlocked competition for strategic and military superiority between the two superpowers. "The decision," said *Pierre E. Trudeau,* "lies essentially in the hands of two men only, one in Washington, the other in Moscow." But this is an oversimplification. The present occupants of the White House and the Kremlin are as much prisoners of their systems' basic rationale and compulsions of their power-elites, as they are captives of their own limitations and prejudices. It is not that they wouldn't want to jump off the tiger's back of the arms race orgy; they can't.

While pacifists cry out their impotent protests, the military all over the world go right on sapping man's limited resources, and drawing the rope of their megaton-biased logic always tighter around his neck. The civilian population of one nation has become the hostage of the nuclear potential of the other, the poor South hostage to the arms race of the North. "The technological imperative is again upon us," implores the U.S. physicist *Hans A. Bethe*. The tautological rule, that to assuage one's own fears requires increased threats of the security of the other, holds sway over all legal and moral considerations.

The second summit meeting between *President Reagan* and *Chairman Gorbachev* in Reykjavik in October 1986 has lightened a faint streak of hope that the two superpowers may yet be able to check their headlong lunge toward the nuclear precipice. Preliminary agreements seem to have been reached on scrapping all medium-range missiles in Europe, limiting intercontinental missiles to one-hundred each in Asia and America, and on 50 percent cuts in all strategic weapons over five years. If these agreements will be ratified by both sides, they will make the threat of war recede and usher in an era based on respect for each other's integrity rather than on conquest and world domination. But whether these rosy perspectives will materialize depends on many imponderables. Past disillusionments counsel caution if not skepticism. Fundamental political and ideological rivalries still divide the world and the two vying systems into profoundly hostile camps. New horror weapons, laser-directed warheads, death rays, satellite-killers, submarine detection systems, and frightful chemical and biological weapons are leaving the production lines or drawing boards every day. The eerie Star Wars weaponry is incontrovertibly calling forth an equally eerie Soviet response. It would be naive to imagine that the involved huge war machineries could be cancelled out at short notice. It has all gone on for so long, as if destiny itself were at work, as if contemporaneous history was propelled by a fateful natural law or a malediction over which man has no control, so that many doubts prevail.

Among the many causes and rationales of the self-perpetuation of the seemingly irreversible trend of the arms race, I think the following play a particularly decisive role.

First, military elites thrive on strained international relations and dangers of war. Peace is a warrior's death. The size of military budgets and the legitimization of the militaries' existence stand in direct relation to their interest in perpetuating real or spurious threats to national security. The Soviet and American military establishments are no exception to this rule.

Second, the Soviet and American planning and R & D establishments have grown into gigantic producers of military wizardry and war-psychosis. With access to almost unlimited resources, and in touch with the power centers, these scientific elites collaborate, on the whole, closely with the schemes, calling for always more deadly and sophisticated weapons, which as *Robert Oppenheimer* once put it, "are technologically too sweet to resist." Arms manufacturers, the huge corporations which pocket billion-dollar defense contracts, also profit more in times when national survival and the international order are at stake than when peace bells are ringing.

Third, the international arms trade is not only a most lucrative endeavor of arms-producing countries, but it is also a factor that helps governments in many Western countries overcome difficult periods of economic recession. According to a study of the U.S. Department of Labor, "a billion dollars of military expenditure creates 76,000 jobs." France and Great Britain are known for treating their arms producing industries with particular favoritism because of their importance for employment and their high profitability. The international arms trade is therefore a factor that propitiates tensions and military conflicts and that would clearly suffer severe setbacks from deescalation and disarmament.

Fourth, the arms race is not only spurred by big power politics in the traditional sense, but by the intrinsic and fundamental differences between the two political and economic systems, engaged in a historical contest for supremacy and survival, a contest also of ideas and basic values that is not going to be over, even if an arms reduction deal prospers. The fact that the leading capitalist nations have more productive, dynamic, and liberal societies will continue to exert tremendous pressures on the Soviet system and make it very difficult for the Soviet leaders to agree to a kind of detente that may look to its Third World allies like a sellout and threaten the pillars of its totalitarian regime at home.

Fifth, a relaxation of the East-West conflict will not do away with the excruciating North-South gap and discrepancies. "Crazy states and movements" will probably continue acting on their own, regardless of Washington's or Moscow's tune, confronting the superpowers with consumed facts, that might tilt "balance of power" quickly into "balance of terror," unhinge agreed upon correlations and equities, and touch off new scrambles for military and strategic superiority.

Sixth, lest the virus of defeatism begins to corrode the fighting morale of their armed forces, Soviet and American military machines uphold the thesis that even a nuclear war, once inevitable, "can and must be won." In case of a major political or strategic setback, the danger that either of the two high military commands may get out of control and feel tempted for "trying to win it" will remain high.

Seventh, both superpowers hold that world peace depends on their strategic and military parity. But since neither can afford to trust the other and neither knows for sure whether he is behind or not, both continue to strive for superiority, impelling the arms race in the way of a "perpetual mobile." Closely related with this is the tautological rationale, "I strike first, says one side, because I am ahead and can win. But I must strike first, decides the other, because I am behind and must impede getting hit first." It is the typical case, as *von Weizsäcker* holds, where "both contenders are chasing a bait tied up in front of their noses that they can never reach, because for both to be superior to each other is impossible."[19]

Eighth, there is a dangerous automatism built-in in the development process of military gadgetry that propels the arms race continuously forward. Every new weapon of one side forces the other to neutralize it with an even mightier response. But military R & D establishments go even further and try to anticipate how the other side will respond to one of their own new weapon systems and how they, on their part, could beat that response. The United States, it was revealed by *Jack Ruina,* a physicist of MIT, in October 1985, is working on offensive arms, which are supposed to neutralize the defensive shield, that the Soviets are introducing to protect themselves from Star Wars. Since both sides feel obliged to act under this compulsion, it is hard to see how this maddening race for the ultimate weapon may be stopped.

Probabilities that the present "no war, no peace" stalemate, highlighted by a criminal squandering of resources for purely destructive ends and awe-inspiring horror scenarios, might be overcome in the near future, are not very high. Bringing the rampant arms race to a halt presupposes a new era of mutual trust, which is conspicuously missing. The Soviet Union would have to renounce on world conquest and on destabilizing the West. The U.S. would have to realize that as a champion of political reaction and of an international order, dividing nations into "have and have not," cannot help lead mankind out of the present chaos and into a brighter future. Whether the pending rounds of the *Reagan-Gorbachev* summit talks will show the necessary insight, and agreement on such far-reaching concessions is an issue open to serious questions.

Erhard Eppler, a leading German social democrat, once postulated that "if peace can no longer be obtained from those at the top, it would have to be imposed upon them from the bottom."[20] But those at the bottom in the North and South alike, who would bear the brunt of sacrifices in a world conflagration, still lack the political muscle needed to accomplish this task.

Contact with a superior civilization would give the forces calling for a stop of the demented confrontation and arms race course a powerful shot in the arm. An intelligent race, thousands and maybe millions of years ahead of us scientifically and socially, is bound to have overcome eons ago the stages of development, marked by intraspecies warfare and near self-extinction, through which our age is passing. This knowledge alone would prove an enormously exciting and rich experience. It may open man's vision to fantastic horizons of organizational harmony and perfection and provide world-conscience with the brutal, eye-opening jolt that may get our civilization, still gripped by blind conflict-fixations, moving toward more peaceful shores.

Chapter 6

World Peace: A Dream Gone Wrong

Perfect peace is achieved in two places only: in the grave and at the typewriter.

—Richard Nixon

What a terrible fix we are in now; peace has been declared.

—Napoleon

It is a basic trait of man to endow his creations with nobler aims and higher objectives than effective means to achieve them, and when failure occurs, to blame his creation rather than himself. The United Nations is a fine example.

When World War II ended, and mankind began taking stock of the deep wounds it had inflicted upon itself, there was hope that with the aid of the newly born United Nations, war, the scourge of humanity, would be deterred in the future, and that world peace would at last become a reality. People were convinced that the nightmare of this conflagration would serve as a permanent exorcising example, and that on the basis of a broad international demonstration of good will, the new world organization would be able to avert war and violence between nations. The spirit pervading the Preamble and the Charter conveyed the sensation that a new hour had struck, and that henceforth, international relations would no longer be marked by the guile of power politics and aggression, allowing law and justice to guide all nations, big and small, to independence, progress, and well being.

Today, more than forty years after its creation, we know that the UN failed to measure up to these lofty expectations. The silent "March of Peace" in New York, on June 26, 1985, in commemoration of the fortieth anniversary of the signature of the UN's Charter, headed by the emissaries of over 150 member states, could not disguise this fact. Nor could it dispel the threat of the nuclear apocalypse that has accompanied us for decades. Peace has been an illusive goal, and war and violence—as we now know too well—have stayed with us all along. Since World War II, as findings of peace researchers corroborate, there were only twenty-six days when no fighting with weapons occurred, and on the average, eleven theaters of war exist in the world on every single day. Between 1945 and 1975, as a study of the Brookings Institute revealed, there were 215 incidents in which the U.S. had used military force as a political instrument, while the Soviet Union had reverted to it on at least seventy occasions.[1] The number of refugees from wars and other violent confrontations since 1945 will soon reach the 20 million mark, the number of the killed already tops World War II casualties. No surprise, considering that half of the world's population languishes under the rule of opprobrious dictatorships.

From the war in Korea to the bloody conflict between Iran and Iraq stretches a long, almost uninterrupted chain of military confrontations, in crass violation of the Charter of the United Nations. Even less successful was the UN with regard to the more than 130 revolutions, coup d'etats and other violent conflicts in Africa, Asia, and Latin America, many of which are still raging on without a UN-sponsored negotiated settlement in sight. The UN was also a flop in relation to the Soviet military interventions in Hungary and Czechslovakia in 1956 and 1968, and more recently in Afghanistan as well as the U.S. military engagements in Vietnam, Cuba, the Dominican Republic, and Grenada. All this shows that nations have continued to wage war and to resort to violence whenever they saw fit to do so, regardless of UN provisions and protests against it. Belligerents and, above all, the aggressors just ignored the world organization, scaling it down to the role of a mere statistician.

This does not mean that the UN was and is a total failure. It is, as *Lester Brown,* put it, "a forum for discussion of international issues" and, as a report of the Club of Rome states, an arena of debate about "mankind's goals and anxieties."[2] The General Assembly dealt repeatedly, though unsuccessfully, with the disarmament issue and the peaceful use of atomic energy, and dispatched UN troops for separating warring parties. The UN also inspired important codes on human rights, supported the struggle of colonial countries for independence, and beat the drums of world conscience on racial equality, as with regard to apartheid in South Africa. Its specialized agencies like FAO, UNIDO, IAEA, and others have come to grips with many of the pressing global problems and are doing a noteworthy job in multilateral development aid. Mammoth conferences, sponsored by the UN, on such topics as the world's population, environment, habitat, science and technology problems, have not only fostered global outlooks, but also established guidelines for important universal action plans in such vital fields as food production and health and technology transfer for development. Other worthy UN activities are geared to further industrialization, as well as education and communications, in the Third World.

The contributions in relief distribution and resettlement of refugees and the resolutions against international terrorism and hijacking of airplanes also belong to its creditable accomplishments, though the latter failed to produce effective results. Of particular significance was the UN Declaration and Program for the Establishment of a New International Economic Order (NIEO), UNESCO's support of a more balanced International Information Order (NIIO), and the UN Treaty on Outer Space, including the moon and other celestial bodies, reserving them only for peaceful purposes.

It is true that most of the UN's declarations and resolutions and charters, as those on Human Rights and on Rights and Duties of States and the Ocean Convention, possess only a declaratory nature, because the UN lacks legal provisions and executive power to enforce them. Many of them are violated and dragged into the mud continuously by unobliging member states. Yet, in spite of its shortcomings, as *Jack Donnelly* pointed out, the United Nations has played and is still playing a very useful role in "standard setting" and in "mobilizing and directing international public opinion on paramount world issues."[3] It symbolizes the unity and common destiny of man, and its Charter, as the Ecuadorian *Leopoldo Benitez* declared to the meeting of ex-presidents of the General Assembly in June 1985, "is still one of the most beautiful documents that has emanated from the human mind."

But as to its peacekeeping mission, the UN's record has been unimpressive. "Safe perhaps," as *Coral Bell* stressed, "what might be called the 'tidying-up phase,' after the resolution of a crisis." *Jeane Kirkpatrick,* the former U.S. Ambassador at the UN, is even more critical. She accused the United Nations as serving "as a battlefield in which conflicts get polarized," fomenting only "strident rhetorics and intransigent postures."

Among the many reasons that explain the UN's poor performance as a guardian of world peace, the following is particularly decisive. From the outset it was not equipped with the authority and legal means that might have enabled its top decision-making bodies, the Security Council and the General Assembly, to live up to this mission. Close to the end of the Second World War, Secretary of State *Cordell Hall* declared to the U.S. Congress that "with the creation of the United Nations there will be no longer need for spheres of interests, for alliances, for balance of power . . . ." This opinion, as *Richard Nixon* puts it, epitomizes "America's naiveté about the nature of the postwar world."[4] This naiveté was quickly shattered by the abrupt end of the wartime U.S.A.-Soviet honeymoon and the eruption of the deep rift between the two countries that has been with us ever since. The world was brutally awakened to the fact that the five victorious powers had reserved to themselves the "Veto Right" in the Security Council and that in absence of a unanimous vote, no sanctions could be applied against violators of the Charter. The Soviet-American confrontation, exploded with unexpected fury, the rugged game of power politics imposed its rules and in no time the UN saw itself transformed in the platform of a grueling wrestling match between the two superpowers. Reluctantly, world opinion accustomed itself to the tiresome routine of reciprocate phillipics, Soviet vetoes, and frustrating tugs-of-war in the Security Council and General Assembly, which demonstrated that underneath all the glittering facade of the UN, the cruel "law of the stronger" was ruling again. The hailed world organization was handcuffed to a meek spectator role, while mankind was irremediably drawn from one major political crisis into another and in the arms bristling stalemate situation of today.

Blame for its failure to prevent these crises, from the Berlin blockade to the Cuban missile crisis, from the wars in Korea and Vietnam to the bloody confrontations and convulsions in the Middle East and in Central and South America, and to serve as an effective mediator of conflict-deescalation, cannot be laid at the doorsteps of the UN building in New York. This responsibility lies with the Big Five and their founding fathers, who never seriously considered relinquishing an iota of their unrestricted sovereign rights and endowing it with the necessary supra-national competence, equal to the task of peace enforcement.

The basic weakness of the UN system cannot be remedied by overcoming "the veil of silence," which allegedly surrounds its positive achievements. Nor will it be surmounted by raising the efficiency of its agencies, augmenting its concerted actions in priority fields or intensifying the relations between the Secretary General and member states, as a critical report suggested. In order to transform the UN into an entity capable of accomplishing the peacekeeping functions consecrated in its charter, more than just cosmetic changes are necessary. It would require a profound overhaul of the Charter's basic provisions, which—in contradiction with the high-sounding ideas and principles of the Preamble—are firmly rooted in the absolutist prerogatives of the "nation state," in arbitrary national sovereignty. The Veto Right would have to be abolished, the number of seats in the Security Council increased, and the obliging character of its resolutions and its power to enforce them upon recalcitrant, peace-violating states greatly enhanced, at the expense of the principle of the nation-state to abide by no higher law than its own.

There is no more urgent need today than to create an international order capable of reversing mankind's drift toward a nuclear cataclysm and of coping with the preposterous inequalities and other chasms dividing the globe. A prerequisite is the gradual evolution of a genuinely supra-national organization, the creation of a world state with a world government, which would supersede the chaotic order of the sovereign state, engaging arbitrarily in warfare and repression of fundamental human rights. Lasting world peace and world government are Siamese twins—one cannot exist without the other. Without the establishment of such a global world order, the curse of war and international violence are bound to remain our civilization's inseparable crosses, and peace will keep eluding us with the perfidy of a beautiful hallucination.

The genius of man is his ability to conceive brilliant ideas well ahead of time, but the tragic irony may well be his incapacity to turn them into reality once the time is ripe. The idea of a world state is a good example. *Marcus Aurelius* already dreamt of a world empire. The French *Abbé St. Pierre* and the British utilitarian *Jeremy Bentham* were both prophets of a world league of nations. At the end of the seventeenth century, the German philosopher *Christian Wolff* coined the term *civitas maxima* and formed the concept that law of the world state is superior to national law. A century later, *Leibnitz, Herder,* and *Kant* laid the foundations of the idea of a great league of nations, encompassing all nations and functioning as the guardian of peace and well-being of all. But neither during the Roman Empire nor in the seventeenth and eighteenth centuries existed conditions for a world state. Today, for the first time in history, compelling political, economic, scientific, and moral reasons speak for the creation of a new global order in the form of a world state as the only real and lasting panacea to mankind's predicaments. Yet as a possible political option, it is no more than a fancy dream.

During the past three decades, many prominent thinkers have come out strongly in favor of world government. *Herman Kahn* pleaded twenty years ago that "the ultimate solution to the Armageddon, Camlan, and 1881—1914 problems is some form of arms control and rule of law, possibly under a world government."[5] The historian *Arnold Toynbee* stated "mankind longs today for a world united in peace and freedom" and he felt hopeful that the scarcity of resources, the population explosion and the threat of nuclear genocide would eventually lead to "the establishment of an effective worldwide government."[6] *Phillip Jessup, Hans Morgenthau,* and *Carl F. von Weizsäcker* have advocated similar ideas.

In my doctoral thesis, "From the League of Nations over the United Nations to a True World Community," presented forty years ago under the impression of World War II, I argued that evolution of international law would necessarily lead to a world state, but that its development, as *Hans Kelsen* had postulated, would have to pass the same stages as national law. The latter evolved starting with rudimentary forms of the judiciary branch. The executive branch followed later, and only in the last stage did codification of common law and the creation of the legislative branch occur.

If this thesis is correct, international law is still in a primitive development stage. The Hague Tribunal exists since 1899, but states rarely submit their disputes to its arbitration, nor do they—in absence of international organs of enforcement—necessarily abide by its decisions. There is still no international executive power capable of bringing an aggressor to his heels, and as to international legislation ruling the relations between states, it is only a fragmentary and weak law, which states violate with impunity whenever it suits their interests. International law is consequently still a very primitive law and considering the lack of international consent, agreement to endow the United Nations with truly supra-national judiciary, executive, and legislative authority that would enable it to impose supreme norms of conduct on nation-states, including the superpowers, will hardly be reached for some time to come.

But the world-state may not come about only in the form of a democratically created world federation. As *von Weizsäcker* has envisioned, it may become reality as a dictatorship in consequence of war. The victor of a final nuclear showdown, either the Soviet Union or the U.S., could use its unchallenged military and scientific superiority to impose on the rest of the globe a single and uniform world order, binding for all nation-states. Maintaining a monopoly of all nuclear and other sophisticated weaponry, the victorious power could outlaw the possession of major weapon systems and armed forces by all other nations, and force upon the rest of the world the political and economic order it may consider the most desirable. *Adolf Hitler,* if luck had given him the atomic bomb and victory in World War II, would have done just that. What a world order imposed by either the U.S. or the Soviet Union would look like is a question I want to address later. Undoubtedly, though, the establishment of a world state in the form of a world dictatorship under the Stars and Stripes or the red flag is not the way in which freedom-loving nations should wish the new world order to come into being.

But what kind of a world-state are we talking about? The prototype I have in mind is a world federal state that may be constituted much in the same way as, for instance, the United States or the Federal Republic of Germany, with central authorities for global management, lawmaking, and jurisdiction and considerable autonomy for the members, safeguarding decentralized execution of policies and cultural identity. What would no longer exist is the sovereignty dogma. It would have a world government, in charge of all global command, control, and coordination functions, a world parliament, to whose legislative corpora would be subjected all other law and a world supreme court, enpowered to settle all disputes emanating from breaches of the world-state's basic charter. Its laws would be binding for all members and its executive bodies would be invested with the authority to thwart aggression, if necessary, by force, and ensure compliance with global guidelines and policies, essential for the peaceful evolution and well-being of the human race.

In *The Third Wave,* author *Alvin Toffler* takes a dim view of a centralized world government, saying that "its thinking is based on simplistic extensions of Second Wave principles," rooted in the industrial revolution.[7] The emergence of a Matrix-System, meshing "different kinds of organizations with common interests," as an Ocean's Matrix, a Space Matrix, or an Energy Matrix, Toffler holds, is a more likely and better alternative. But mankind's most dramatic issues, the East-West confrontation, the North-South gap, the tottering world economy, and the nuclear threat, will not be overcome by such Matrix Systems. It is the arbitrary sovereign nation state that must wither away and dissolve into the larger unit of a World Federal State, just as tribes merged in city-states and dukedoms, and these eventually merged into nations.

The epoch-making meaning of the creation of a world-state is that it would permit mankind to outlaw war and to purge aggression and violence from international relations. The gigantic stockpiles for nuclear, chemical, biological, and conventional warfare could be destroyed, armies disbanded, and military scientists and generals reassigned to peaceful jobs. The astronomical military expenditures, which reached $800 billion in 1986, could then be destined to more constructive human ends. The danger of Star Wars and Nuclear Winter, so everpresent today, would become nightmares of the past, never to return again. The elimination of war, as the most hideous scourge of the human race throughout its history, and the realization of "eternal peace," which *Kant* dreamt about two-hundred years ago, would finally become reality.

Organized in a world-state, mankind would also be able to tackle in a much more effective way the heavy economic and social problem load that keeps undermining and destabilizing the present world order. It would be no minor task. Not even a world government could turn our disjointed and unequally endowed family of nations into a prosperous and harmonious world community just in a couple of decades. But I believe that with the determined cooperation of all advanced countries, buttressed up by the world's material, technical and financial resources, and with the aid of genuine efforts of developing countries, formidable strides with regard to the paramount social and economic ills of the poor half of the globe could be achieved. No longer hamstrung by parochial rivalries and suicidal confrontations, mankind could then, as a whole, embark on a full-scale and long-term attack on the gruesome inequities and sufferings, still besetting the greater part of the world's population and put its fabulous scientific and technological potential entirely at the service of the inalienable right of all people to live a life free of want and oppression. Internationalization of the world's strategic, natural, and financial resources, with the aim being to propitiate rational exploitation and growth, and global planning, in order to ensure balanced economic performance and distribution patterns, would facilitate the task. World citizenship, unrestricted world travel, and concerted efforts to give man everywhere access to the marvelous communication facilities already within our reach, would also help to create a new world conscience, cemented in the concept of the political, economic, and moral oneness of the human race. In a century or two, mankind will have progressed a long way along the road of eradicating most of today's dire problem-issues. So conceived, the world-state would mark the final departure point of our species' primitive organizational phase, characterized by disunity, perpetual conflict and war, and the dawn of a new historical age under the token of peace, the rule of reason, and genuine brothership of man.

All this seems like a fantastic glimpse of Alice in the wonderland. The world federal state, for all the urgency to do away with the dangerous and unjust nation-state system, is still a utopia, " . . . so remote," as *von Weizsäcker* ruefully admits, "that nobody of us knows how and when it could be achieved."[8]

Not all opinions, however, are so pessimistic. Years ago, *Herman Kahn* stressed. "It is clear that requirements of preserving peace and the problems of arms control, the environment and economic relations, as well as many law and order issues, all create great pressures toward peaceful evolution to world federal government."[9] Growing economic interdependence and the advances of science and technology, busily promoting what *Toynbee* used to call "the annihilation of space," are tearing down geographic and legal barriers and increasing the need for international collaboration. Steeped up global distribution of goods and mass dissemination of news, ideas, and symbols are also contributing to the evolution of universal behavior patterns and expectations and the creation of a nascent universal consciousness and culture. Even the United Nations, in spite of its blatant weaknesses, may be seen as the embryonic stage of a future world federal state, and its covenant, resolutions, and declarations may be regarded as forerunners of genuine world law. All these facts and trends, taken together, amount to an impressive current that points toward the eventual implantation of a truly supra-national global order.

Nevertheless, for all practical political purposes and regardless of the blessings it might bring mankind, for the time being, the creation of a world federal state remains a beautiful but utopian goal, mainly because of the following five reasons.

The East-West Conflict. As long as the world is divided in two hostile camps, vying with each other for world supremacy, trust and will on both sides to bury political and ideological differences and to join hands for building a new, just and peaceful world order, the prerequisites for any consensus on transforming today's sytem of the United Nations into an effective supra-national organization, are clearly missing. The cruel rationales nourishing the conflict between the two superpowers condemn any initiative in this direction to immediate failure.

The Durability of the Nation-State and Sovereignty Dogma. The maxim that the state is the ultimate judge of its actions and the principles of "non-intervention" and "non-interference" in another country's affairs are still guiding rules of international relations, jealously upheld especially by smaller nations. Nationalism, as *George F. Kennan* pointed out, "is still a tremendous political-emotional force, that crafty statesmen and politicians keep exploiting to the hilt."[10] But the problem is more than nationalist chauvinism. Even in industrialized countries, the nation-state keeps accruing power as crisis manager and promoter of giant defense, space, and social programs. In many Third World countries, too, the role of the state is augmenting as a prime motor of national development. It is therefore highly unlikely that we shall witness the demise of the nation-state and the sovereignty dogma in this age.

The Interest of Power Elite in the Status Quo. Power elites everywhere, in rich or poor countries, in Communist or non-Communist regimes, are profoundly conservative groups. They cling to the wealth, power, and status they have acquired, and abhor radical changes that may force them to relinquish it all. Political elites are much more interested in preserving their privileged positions than in furthering solutions to the world's gravest problems. For similar reasons, why should the world's most powerful economic and military elites, who profit most from our dangerous stalemate, want to disavow the nation-state and give their support to a world federation that may quickly curtail their rank and arbitrary powers, and strip them of their arms and medals? Class and caste interests are bound to present stern opposition to the global interests and idea of a world-state.

Lack of Consent Regarding the New World Order. The creation of a world federal state would necessarily presuppose a basic international consensus on the kind of world order that should be established. There would have to be general agreement on how to do away with war for good, of how to overcome the excruciating North-South disparities, and of how to address the many other problems-issues besetting mankind. Especially there would have to be some consent with regard to the political color of the new order, whether it should be tailored in accordance with the Western capitalist or Communist model or a mixture between both. But no common blueprint for such a grand-scale enterprise, reconciling the parochial proposals of each side, so far exists and it is likely, as the global conflict keeps growing, that no agreement on such a blueprint will be reached in the near future.

The Missing External Threat. In one of his writings in 1928, *André Malraux* had world leaders plot a world government movement on the grounds of an alleged pending invasion from the moon.[11] The nations united, especially when it turned out that the threat was real. In Geneva, *President Reagan* is supposed to have said to *Chairman Gorbachev,* "If extraterrestrials would threaten Earth, Soviet and American forces would quickly ally to repel the attack." However, no such common enemy, which might serve as a powerful stimulus for peace and cooperation between the U.S. and the Soviet Union, and for the rapid transformation of the 155 nation-states in a world federal state, exists. Even if it existed, international relations would probably quickly relapse in the "status quo ante," once the external threat had been defeated.

Real probabilities that the precarious UN system may be transformed gradually and by a majority consent in something resembling a world federal state in the foreseeable future are therefore practically nonexistent. There is, however, as I pointed out earlier, an alternative possibility by which we might get catapulted into a world-state rather quickly, though not unscathed; and as of the time of this writing, this alternative does not look unlikely at all. *McGeorge Bundy* envisioned that by 2024, after the ravages of a nuclear war and famine and some 65 million people are killed, the survivors would agree to the establishment of a supreme world authority, which would control nuclear weapons and population growth, among other things. In the doctoral thesis, presented in 1948, my thoughts ran along similar lines. Much to the displeasure of my counseling professor, I argued that our world would only be ready to relinquish the absolutist sovereign state and accept the creation of a true supra-national authority, after going through the macabre experience of a third World War. The fact that the League of Nations had been the result of the first World War, and that the United Nations became a reality only after World War II, provided convincing examples. Referring himself to the "unthinkable" happening, the German writer *Gerd von Hassler* arrived reluctantly at the conclusion, "Three generations hence, even its horrors could turn into a paling memory. We should not deny the existing chance that people will live then in a less crowded and therefore less aggressive and better world than today."[12]

The mere thought that future history may bear out this thesis is chilling and appears totally cynical. What a terrible price to pay for world government, even if it would bring us Kant's eternal peace! Chances that the raging power struggle between the U.S. and the Soviet Union will plunge humankind in a nuclear furnace are great, and neither for the doomed to perish in it, nor for the few survivors will it be any solace to know that from its radioactive debris, a more rational world order will arise. Scientists speak of the fundamental unpredictability of nuclear war and that it spreads uncertainties like a wheel of firecrackers. The danger is that they will keep alluding to these uncertainties, until someday somebody may feel tempted to find out just what they are.

But there is still another school of thought, making rounds in some scientific circles, whether a peaceful world-state would not inevitably lead to biological degeneration and cultural stagnation? Without the continuous spur of war, hold the epigones of *Oswald Spengler's* pessimism, nothing can prevent civilization from falling prey to progressive decadence and senility.

I think this fear is unwarranted. The creation of a world federal state would help unwind enormous potentialities for the full mobilization of man's constructive resourcefulness and energy. To remodel the world in the image of a rational, just, and peaceful global order should be a colossal challenge, keeping his imagination and knack for accomplishing big things busy for a couple of centuries. Space exploration and exploitation should also give his quest for adventure infinite scope. Once mankind has learned how to live in peace and share its wealth and knowledge for the well-being and progress of all people, learning how to impede atrophy of man's creative springs and a possible culture-crisis should be an easy job.

Humanity's advance toward the world federal state is still a long climb up a steep mountain. Again, the extraterrestrial connection may provide us with meaningful insights. Much older and further advanced extraterrestrial civilizations must have established a stable peaceful global order on their celestial habitat a long time ago. Probably run by unimaginably smart computerized systems and maybe genetically programmed robot machines, their organizational system is bound to function with the accuracy and harmonious hum of an atomic clock, Intra-species violence and warfare would have ceased to be a topic of such intelligent beings eons ago.

But for us, making contact with such a perfectly stabilized and peaceful order would be an enormously stimulating experience. At first we may gasp in awe at a system that has forgotten the meaning of war and destructive fanaticisms. At the same time, it would intrigue us enormously to find out whether there is a road leading to a peaceful global order without the need of a purgatorial genocide. If there was one in the extraterrestrials' long past—though we wouldn't be able to copy it—it may teach us a lesson. It may dawn on us that we can't go on forever trying to reach genuine peace either via a truce of terror or the petrification of an unsustainable status quo. The reciprocal reduction of strategic U.S. and Soviet missiles will probably diminish the danger of all-out war, but it will not usher in peace. I am afraid that as long as the words "liberty" and "equality" are not undressed of the farcical meaning, which they have for four-fifths of mankind, genuine peace will remain an illusionary goal, and for many, an unworthy alternative to violence. Contact with highly developed extraterrestrial intelligence will also be a shocking experience. Our future may depend a great deal on our capacity to absorb it.

Chapter 7

The Quicksands of Revolution

The people who do the most harm are the people who try to do most good.

—Oscar Wilde

Two decades ago, really throughout the sixties, there was politically something exciting in the air. A revolutionary tide seemed to be on the loose, sweeping through developed, as well as underdeveloped, countries. It questioned the very basis of existing structures and values and threatened to plunge Western societies into a turmoil of social confrontation. It was mainly the revolt of the younger generations who felt repulsed by the gruesome contradictions underlying the existing order, peace turned into a chimera, rampaging technology, the growing gap between "have" and "have not" nations, the materialist-consumerist culture patterns, and the manifest incapacity of the establishments to cope with them. A revolutionary zeal was aglow, especially in academic quarters, leading in countries such as the U.S., France, and Germany to violent student demonstrations, and in many Third World nations, especially in Latin America, to bloody guerrilla warfare. One basic credo united them all—the corrupt and oppressive existing order should be overthrown, if necessary via violence and civil war. *Marcuse, Fannon,* and *Marx* were the mentors; *Ché Guevara,* the rousing symbol.

Now this revolutionary fire is burning with a much dimmer glow. Diehard groups such as the *Black Panthers* in the U.S., and the *Baader-Meinhof* groups in the Federal Republic of Germany have ceased to exist. The *Red Brigades,* once the terror of Italy, have come apart by the seams after the turnabout of their top leaders. Even the revolutionary struggle in Latin America, which *Ché Guevara* had hoped to turn into another Vietnam, has lost most of its initial impetus and seems to be petering out slowly. In Argentina and Uruguay, the military disposed with unprecedented brutality of the "Montoneros" and "Tupamarus," and *Raul Sendic,* the Tupamaru's leader, proclaimed upon his release in 1985 from thirteen years in prison that "henceforth

72

the movement would incorporate itself to the (peaceful) political struggle."[1] In Colombia, too, a country with a particularly long record of violence and subversion, the pro-Soviet FARC, led by legendary "Sure Shot," accepted the olive branch of internal peace, tended by former *President Belisario Betancourt,* and to participate in the elections of 1986 as a political party. Peace talks have also been started between the government of El Salvador and the guerrilla coalition in 1984, which may yet lead to pacification of this small, tormented country.

The frightening era of revolutionary strife and violence is, of course, still far from over. The assassinations of prominent political leaders, as the recent murders of *Indira Gandhi* and *Olof Palme* have demonstrated, is a profession still practiced with fanaticism and skill. Political terrorism in the form of *Carlos's* wanton bombings, or kidnapings of planes or taking of hostages in the fashion of *Abu Nidal,* or any of the ultra-radical factions in Lebanon, as a tool of political blackmailing, remains a successful métier. The outlawed IRA and ETA armies in Northern Ireland and Spain still rely on terror for furthering their aims, and the recent assassinations of members of the German and French industrial, political, and military establishments show that minuscule European "desperados" have still not learned the lesson from the failures of their forerunners. In Colombia, too, a few ultra-leftist guerrillas, especially the "M-19" movement, and in Peru the belligerent "Shining Path" movement, still hitch their fortunes on the strength of their guns in vain bids for power. In other parts of the world, *Arafat's* once powerful PLO, weakened by internal divisions and military setbacks, is still alive, while the Islamic Revolution of 1978-79, proclaimed by *Ayatollah Khomeini,* is spending itself in the brutish war with Iraq. In South Africa, the opposition of the large Negro majority against apartheid may yet explode in a bloody revolution.

The reasons why revolution in some countries is still a battle cry are easily stated. With regard to the roots of the powder keg situation in Central America, the report of the *Kissinger Commission* admitted that "Discontent is real and generalized and that, for large numbers of the population, living conditions are miserable: just as Nicaragua was mature for its revolution, in the same way, conditions inviting a revolution are present in the whole region."[2] As to the situation in Peru, *Cynthia McClintock,* who did extensive research on the "Shining Path," states, "People in the southern highlands earn little, die young, are mostly illiterate, and usually exist without human services."[3] Quoting sources showing that the highland farm incomes had dropped in 1981 to "below fifty dollars a year per capita," and that individuals in these areas "were apparently consuming as little as 420 calories a

73

day," this American author counts the population of the southern highlands of Peru, where the Shining Path movement runs strongest, to the poorest of the world. The Sandinistas, too, had a powerful cause on their side, fighting against *Somoza's* mercenary army and a political system guilty of forty years of ruthless tyranny and corruption. The same lance may be broken for *Arafat's* PLO, fighting for the right of Palestinians to return to their homeland and the right of self-determination, though this does not justify the bloody terrorist trail sown by its fighters. Even *Khomeini's* unsavory Iranian Revolution had history on its side for everthrowing the reactionary and corrupt police state of the *Shah,* as did *Fidel Castro's* popular uprising against the ugly dictatorship of *General Batista* thirty years ago.

But these examples do not invalidate the fact that, by and large, faith in the infallibility of violence and revolution is slowly waning and giving way to more sober reflections about adequate ways for furthering political and social change. Where powerful nationalist, racial, or religious motivations play a predominant role, as in the case of the Palestinians, the South African blacks, and the warring parties in Lebanon, street violence, terrorism, and civil-war–like confrontation will continue reaping a gory harvest for some time to come. But where these factors are missing or where dogmatic ideologists have vainly tried to impose a revolutionary panacea from above, many of the realistic thinking revolutionaries have begun to sound the trumpet of retreat. There have been too many mistakes, aberrations, failures, false messiahs, useless sacrifices, and frustrating defeats. The isolated terrorists acts in France and Germany only help to underscore the political impotence of a few handfuls of professional conspirators and incorrigible system-haters. Political terrorism, in the form of the kidnapings in Lebanon, as a means to extort from Israel and the U.S. major concessions, are also just desperado acts. Ironically enough, as Iranscam has illustrated, they so far yield tangible results.

In other parts of the world, the revolutionary myth, which missionary leaders such as *Malcolm X, Ché Guevara* or *Camilo Torres* have been able to kindle, has long since paled. This holds true even for many underdeveloped countries, where social inequalities are high, and discontent with the status quo is acute. In South America, more than twenty-five years of urban and rural guerrilla warfare in all countries except Chile and Paraguay, have left a trail of singular failures. Faced by this disappointing record, leaders of several insurgent movements allegedly met in September 1983, creating "a continental alliance" with the aim of "rescuing the revolutionary flag of Ché Guevara and placing Latin America on a war footing against *Yanqui* Imperialism."[4] Plans to organize a revolutionary "Bolivarian Army" with the participation of guerrilla groups of several countries were revived in fall 1985, with no practical results to date. The return to legal political life of the "Tupamaros" in Uruguay and the "FARC" in Colombia, the truce declared by the "Tupac Amaru" movement in Peru, and the setbacks suffered by the "M-19" in Colombia, but above all, the fact that weariness and rejection of sterile violence and useless sacrifices are growing everywhere, are likely to condemn these plans to the same fate experienced by *Ché Guevara* twenty years ago in Bolivia.

Frantz Fanon, who glorified violence as a holy means of the anti-colonial revolt, and *Ché Guevara,* who dreamt of giving U.S. imperialism the coup de grace by stirring up in Latin America a revolutionary inferno, would feel disheartened over the reversals suffered by the revolutionary struggle in many countries and the notable decline of appeal of revolutionary ideals. The leaders of the student revolts in many Western nations in the late sixties, too, must feel disenchanted because of the conspicuous trend away from radical thought at university campuses, and the boom of hedonism and the astonishing resurgence of traditional values in today's young generation.

But, alas, is it not maybe one of the paradoxical facets of our time that this protest and challenge of some of our nation's finest youth, its penetrating attack at basic contradictions of Western society and the existing world order, has faded away without leaving a trace on man's collective conscience? Have the sacrifices of thousands of young idealists, slaughtered in battle or in dungeons, left no impact whatsoever on the fossilized power structures in Latin America and in other continents? Maybe we shouldn't be excessively happy that the student demonstrations of twenty years ago didn't entail constructive consequences, and that they were subdued more by the clubs and guns of the repressive organs than by the force of the ideas of the ruling establishments. Not everything in the chaotic, partly naive, but still idealistic analysis of leaders such as *Angela Davis* and *Eldridge Cleaver* in the U.S., *Rudi Dutschke* in Germany, and *Bendit Kohn* in France, was outright garbage. They expressed only the protest of the young academic generation, bent to change the rules of the game, that had bequeathed humanity only deep political antagonisms, new pariah nations, technological nightmares, cultural degradation, and prospects of nuclear extinction. "The movement of the left during the later sixties," admits *Carl F. von Weizsäcker,* "was an angry-optimistic anticipation of the crisis facing mankind" ever since.[5] In the wake of the new conservative wave, many academic quarters have stopped being fountains of electrifying political thought and debate—the grand social vision has been lost.

But let us turn now to some of the deeper reasons for the receding revolutionary currents in many parts of the world. The most important is, no doubt, that the path of guerrilla warfare and terrorist violence, which was supposed to catapult their leaders into power, has proven costly, futile, and erroneous. Another one is, I think, that in countries where revolutionary groups and movements have succeeded escalating to power, their performance has been anything but convincing, making many believers apostatize the violent cause.

A few examples help illustrate the first point. The "Tupamarus" in Uruguay, and the "Montoneros" and "ERP" in Argentina met total defeat not only because of the ruthlessness of the repressive apparatuses, but mainly because they failed to rally massive popular support. *Ché Guevara's* ill-fated odyssey in Bolivia was doomed to failure for the same reason. The stoic and suspicious highland Indian remained immune to *Ché's* Marxist proselytism. His military and political defeat twenty years ago only proved, as *Regis Debray* belatedly admitted, that the "theory of the focus," developed by *Ché Guevara* and *Fidel Castro,* and embraced by many other guerrilla leaders as a tactical bible, was but a voluntarist thesis that did not stand up to reality. Revolutionary idealists as *Puente Uceda* in Peru, the priest *Camilo Torres* in Colombia, and *Turcios Lima* in Guatemala, learned the fallaciousness of this naive thesis, and paid for the mistake with their lives.

Fanatic guerrilla movements, as the "M-19" in Colombia and the "Shining Path" in Peru, may still hang on to a marginal existence for years, but their one-track doctrinairism and proverbial incapacity to evolve into broad, popular political movements condemn them necessarily to a quixotic pattern of "hit and run" *bandolerismo*—painfully experienced by *Ché Guevara* in Bolivia. At their end beckons only total annihilation, or surrender, or blending in time into a peaceful political force, the alternative chosen by the Colombian FARC. It is true that the "M-19" movement has some roots in rural and urban substrata and that "Shining Path" has gained support from the poverty-stricken peasants of the Peruvian southern highlands. But their unprecedented brutalities, for instance, the suicidal attack of the M-19 on the Palace of Justice in Bogota in 1985 and the many peasant massacres in Peru as well as their rigid advocacy of a Marxist or Maoist dictatorship, disqualify both movements in the eyes of the majority of the people in both countries as viable political options.

Ulrike Meinhof and *Andreas Baader,* cofounders of the German Red Army fraction, committed suicide in prison in 1978, supposedly in an act of protest, to rouse comrades to renewed militancy and revolutionary activities. Closer to truth is probably that both had realized the hopelessness of their struggle, and that terrorism had brought the RAF total discredit and alienation from the great bulk of the German people. The disintegration of the Italian Red Brigades in the early eighties is no less revealing. One of their leaders, *Antonio Savasta,* and three other members, who had kidnapped the American *General James L. Dozier,* advised their comrades in a joint declaration from jail "to abandon the armed struggle, because it resulted in organizational and factual failure."[6] Early in 1983, the collective of the Red Brigade members held in the prison of Palmi, where their top leader *Renato Curcio* serves a life term, caused an éclat with another declaration, in which they subjected ten years of armed revolutionary struggle to a critical review, acknowledging "that none of the organizations succeeded in making the great leap forward; and that it is necessary to admit it, and to renounce on this form of struggle." Clearly the brain trust of the Red Brigades, once Europe's most feared terrorist organization, had become convinced that violence and terrorism were inadequate means for disrupting state institutions and overthrowing capitalism. In the U.S., too, the Black Panthers and the Symbionese Liberation Army, which had kidnapped and recruited *Patty Hearst,* have folded up and now belong to the past. After living for years in several Communist countries, *Eldridge Cleaver,* the Black Panther leader, abjured upon his return to the U.S. in 1975 to the use of violence and declared that, "After living for years under dictatorships, I have more balanced views of what is happening in the world."

With regard to my second point, let us scan four revolutionary processes in Cuba, Chile, Peru and Nicaragua, each of which has received, and partly is still receiving much world attention, and that have garnered much admiration on the part of radicals the world over.

The case of the *Cuban Revolution* is easily stated. It is the case of a patriotic, antidictatorial, and anti-communist movement that, when *Fidel Castro* gained power in 1959, had everything on its side—popular support and worldwide acclaim. But by turning Communist, and using the island as a springboard for continental rebellion only ninety miles from the shores of the U.S., its libertarian essence was betrayed and basic promises remained unfulfilled.

Its practical results are well-known. The Cuban people have been forced to accept the excellencies of the totalitarian state, one party rule, secret police terror, and thought control. After almost thirty years of wanton socialization, its once quite buoyant economy agonizes and is still unable to feed its people. A stringent ration card system and billions of dollars of Soviet aid illustrate the economic plight and the bonds of dependence characterizing the country.

After its ouster from the OAS, Cuba slumped regionally almost to total isolation from the subcontinent, facing for years a hostile ostracism because of its aid to subversion, a situation it is only beginning to overcome. On the international scene, Cuba has been transformed into a pawn of the Soviet Union's worldwide schemes. During the missile crisis in 1962, it almost provided the spark to nuclear holocaust. Since then, its people chafe under the burden of a stifling war economy, engagements of its armed forces in the Congo and Ethiopia, and the hardships of an inefficiently run economy.

It is true that the regime has done much for education, health care, and sports and that wealth and income have been more equitably distributed. But the overall record of the Cuban Revolution has been disappointing. Soviet stylized economic and cultural patterns have chiseled deep inroads in the Cuban soul. Ché Guevara's "new man" has remained a utopia. Cuba's cultural life, rocked by scandalous mock trials, no longer evokes jubilant throngs in Montmartre and Greenwich Village.

Except for the privileged members of the "new class," life for the average Cuban is but a terrible drudgery, a long list of sacrifices asked in the name of the Revolution, unkept promises, and no hope for change. In 1980, more than 10,000 Cubans stormed the Peruvian embassy within twenty-four hours, seeking political asylum; the following year, 120,000 Cubans scrambled to freedom via a sea operation between Port Mariel and Miami. These two events cast a telling light on what the majority of people think of the Castro regime.

Even *Fidel Castro* has begun to recant. In an interview granted to journalists of the *Washington Post* on January 30, 1985, he admitted that inexperience and romanticism had led the Cuban leadership to many mistakes for which the country had to pay dearly.[7] He has discovered merits of the market economy and is flirting with the idea of traveling to Washington for a peace talk with President Reagan. The realities are catching up with the scintillating but costly bunglings of the Cuban Revolution.

President Salvador Allende's ill-fated "Chilean Experiment" is another good example of revolutionary charlatanry. For a brief spell, *Allende's* "peaceful road to socialism" signaled an inspiring message for progressive leftist Latin Americans, anxiously on the lookout for an attractive alternative to the violent Cuban antithesis. But it soon turned out that the model was vitiated by the deadly virus of ultraradicalism from the start.

Under the luring label of "peaceful and democratic transformation" within the constitutional order and liberal traditions, a system of social justice, general welfare, and internal peace was to be achieved. But what were the real accomplishments? Lusty government spending, the precipitated land reform, and discriminate nationalization and socialization programs soon plunged the Chilean economy into anarchy. Illegal seizures of farms and factories, the exodus of thousands of qualified entrepreneurs, and dwindling foreign resources aggravated the situation. Before long a skyrocketing inflation rate, lack of basic supplies, capital flight, and a burgeoning black market highlighted the deep malaise of the Chilean economy, long before the much publicized "marches of the casseroles," and the fatal strikes of the transportation workers.

Chilean jails remained free of political prisoners and opposition parties, and media were allowed to operate freely in a queer melange of libertinage and subversion. But the practice of the government to bypass Congress and to rule by decree, its tolerance of transgressions of law and order by extremist groups, and its clearly Marxist-inspired philosophy and action plans soon escalated the ensuing class conflict to a highly explosive confrontation. The close ties of the Allende government with Cuba, its professed friendship with the Soviet Union, the hateful anti-American campaigns and the covert endorsement of plans to subvert the military and prepare an armed worker's militia all helped conjure up the nemesis of an imminent revolutionary coup d'etat with the aim of establishing a Marxist-leninist dictatorship. By summer 1973, effective government had ceased to exist. When the military stepped in, Chile stood on the verge of civil war.

Allende's "peaceful road to socialism" had only been a masquerade. As he admitted to *Regis Debray,* his differences with Fidel Castro and Ché Guevara were only of a tactical nature. The vows about preserving representative democracy were just a bait to get time and to placate public opinion abroad. The true aim was to destroy bit by bit the class enemy, and to enthrone on the back of its corpse a system much in the style of the Cuban Revolution.

It is a much cherished myth that Chilean reaction and US. imperialism are to blame for the tragic end of *Allende's* experiment. Blinded by the fanciful glory of the Cuban Revolution, he wanted to settle for nothing less. That the menaced bourgeoisie, the landowners, the military, and their foreign allies would defend themselves should have been expected. "It is little statesmenlike," states *Jean F. Revel,* "to blame one's own failures on the lack of support from one's enemies, whom one has pledged to destroy."[8]

The bungled *Allende* experiment bequeathed the Chilean people, in the wake of the takeover of *General Pinochet's* distasteful dictatorship, a terrible bloodbath and a thumping defeat of democracy. It amounted to nothing more than a costly and very disappointing revolutionary pipe dream, misguided from the first day to the last, and at a political and human price that generations will remember.

The third example is the revolutionary debacle in Peru. In 1968, a group of generals under the leadership of *Gen. Juan Velasco Alvarado* seized power and initiated with the aid of leftist intellectuals a revolutionary process with strong anti-capitalist, anti-imperialist, and pro-socialist traits, which stirred the nation and the continent. The sweeping reforms, contemplated in the long secret "Plan Inca," called for a profound overhauling of the traditional Peruvian society and economy.

During the following years a radical land reform was introduced; basic industries such as oil, mining, and fishing were nationalized, and an extravagant system of co-management of workers was imposed on remaining private industry. The national press was seized and assigned to different strata of the population. The aim was clearly to destroy the traditional bipedal capitalist-feudalist society, and its dependence on the U.S., and create a juster society along socialist lines. Little wonder that the Latin American left, with *Fidel Castro* leading the chorus, soon hailed *Gen. Velasco Alvarado* as a great revolutionary leader, and that before long Soviet money, experts, and weaponry began pouring into the country.

Where did the Peruvian Revolution lead up to? By 1975, it was clear that it had failed its high-sounding goals, and that it was in deep trouble. Agricultural production was down and private enterprise agonizing. National reserves had been squandered in overly ambitious investment projects and in huge arms purchases, and foreign resources were drying up. From 1975 to 1978, GNP decreased in real terms, and in 1978, minimum salary in real terms was only 50 percent of that in 1973, as successive devaluations and a virulent inflation rate hit the country. Even after the ouster of *General Velasco* in 1976, and the enactment of an emergency austerity program, the Peruvian economy kept tottering on the brink of collapse. What was worse, real standards of living of low income groups had seriously deteriorated, farm incomes and food consumption in the highlands had decreased, and ambitious expansion programs of education and health services had to be paralyzed.

As a consequence, opposition to the military revolutionary regime soon reached explosive levels. When Peru returned to democracy in 1980, the elected president, *F. Belaunde Terry,* received the country on the verge of economic bankruptcy and social dissolution. The revolutionary romanticists, this time led by military boots, had had another chance to wield power and to ruin a country. However, apart from years of sacrifices and frustrations, *General Velasco's* revolution handed Peruvian people a legacy with a dangerous, delayed effect. The politicization and radicalization promoted by his improvised reforms set the stage for the eruption of the new brand of terrorism and violence initiated by the "Shining Path" movement in the early eighties.

The case of the *Sandinista Revolution* in Nicaragua, the last case described here, has been transformed by a tragic aberrational process into a mockery, and the "number one trouble spot" in the Western hemisphere is still unrolling before our eyes with no happy end in sight. When the Sandinistas toppled the corrupt and hated *Somoza* regime in July 1979, they had the sympathy of the overwhelming majority of the Nicaraguan people and world opinion on their side. The promise of the Revolutionary Patriotic Front to install a democratic system based on the principles of a mixed economy, political pluralism, and non-alignment, seemed to inaugurate at last a period of genuine peace and progress for the 2.9 million Nicaraguans.

But as soon as the Comandantes established themselves in power, they pulled off their democratic masks and began laying the groundwork for the creation of a totalitarian regime. Government policy soon acquired a distinct anti-capitalist and anti-U.S. slant and began harassing private enterprise and foreign investors. The enormous Somoza holdings, together with banking and foreign commerce, were nationalized, television and most of radio socialized, and a stiff censure imposed on the independent press. Democratic opposition parties began to be subjected to state pressures and chicaneries and were soon stripped of almost all their political significance. The Catholic Church and other religious congregations also soon began to feel the wrath of the new leaders in Managua. In the meantime, anti-liberal and anti-U.S. frenzy was whipped up in the masses, while thousands of educators, doctors, and military advisers started pouring in from Cuba together with shiploads of modern armament from Soviet bloc countries. Aid to the guerrilla movement in El Salvador, and to other revolutionary groups in Central America became Nicaragua's covert policy.

What are the practical results of the much heralded Sandinista Revolution? Seven years after its triumph, its overall situation couldn't be bleaker. Democratic opposition has been stifled and the ultra-radical *Daniel Ortega* swept into power. Many outstanding former collaborators and sympathizers have turned their back on the Sandinista government and are actively engaged in overthrowing it. For years a civil war conflict has raged on with the contras, overtly aided by Washington's official policy not to tolerate "hostile Communist colonies in the Americas." On numerous occasions, the strained relations of Nicaragua with its neighbors, Honduras and Costa Rica, have almost escalated into war. The possibility of a U.S. military intervention remains acute, a scenerio which has accelerated Managua's drift into the Soviet-Cuban camp and its dependence on Soviet aid and modern arms. All this underlines the fact that Nicaragua has transformed itself into an acute trouble spot of international dimensions that seriously menaces peace in the hemisphere.

No wonder that Nicaragua's economy has been slipping steadily, hamstrung by bureaucratic inefficiency, lack of private investment, exodus of capital and technicians, speculation and sabotage, but above all, by the exorbitant military expenditures, reaching 40 percent of the budget in 1985. High inflation and unemployment rates, the official admission that "production has fallen to the level of 1977," a severe drop of exports and real wages and its soaring foreign debt, forced the Managua government already in 1981 to declare a state of economic emergency and introduce a rationing system.

The final chapters of the "Sandinista Revolution" have not yet been written. Whether in the aftermath of Iranscam, U.S. aid to the contras will suffer curtailments, giving the Managua hardliners a respite, or whether Washington will eventually be forced to intervene militarily to stamp out this pro-Communist beachhead in its own strategic backyard, still remain options of the future. But one conclusion is evident. The passions of ideological fanaticism have driven the Sandinista leaders and Nicaragua into a pitiful dead-end situation, that makes true development and progress impossible. An imaginative participatory and pluralist democracy would have been a blessing for the downtrodden Nicaraguan masses. The replacement of a reactionary tyranny with a regime moving blindfolded toward a Marxist-Leninist dictatorship is bound to bring them only more hardships, bloodshed, and disillusionments.

A basic common pattern distinguishes the described four revolutionary examples that led them irremediably to their early demise or to their respective present dilemma.

First, they are marked by a rejection of representative democracy and the belief that political and ideological pluralism is incompatible with true national development and independence. This explains the inherent drift toward an always more authoritarian, if not totalitarian, regime.

Second, they are intrinsically opposed to capitalism, private initiative, and the market forces, hinging their hopes primarily at the bloated state economy and global planning.

Third, their orientation is basically anti-American and pro-Soviet, which inserts them necessarily in the global struggle between the two superpowers.

Fourth, they refuse to limit the endeavor to the national level and have an innate tendency to project the revolution continentally.

By clinging doggedly to this dogmatic revolutionary strategy, the inevitable consequences were economic disaster, violent internal confrontations, and serious conflicts at the external front, which either doomed the experiment to failure or brought it very near to it. That the Cuban Revolution has survived is probably mainly due to its sui generis condition as an island. Whether the Sandinista leaders will be as lucky is a big question mark. But there is another, more fundamental fact that the revolutionary leaders, in all four cases, have chosen and still choose to ignore. The goal of an economy that leaves no scope for private initiative and of a regimented, basically unfree society, in which everyone is told what to think, is out of touch with the political and cultural mainstream on the continent.

This is also one of the main reasons why probabilities that the guerrilla struggle, still going on in a few Latin American countries, will be successful are extremely small. Evidently the existence of objective conditions as poverty and social injustice is not enough to spark a broad irresistible revolutionary tide, nor are there viable alternative motivations at hand that may help to bridge the isolation in which most guerrillas operate in the near future. For the diehards, unable to see the light, self-immolation is the only way out from this cul-de-sac situation.

In 1963, almost twenty-five years ago, I submitted to *Ché Guevara* an essay on the basic fallacies of extrapolating the model of the Cuban Revolution for other Latin American countries, and the convenience to broaden its democratic base and normalize its strained relations with its mighty neighbor. He was quick in rejecting all my arguments. But history proved him wrong only four years later in Bolivia, and *Fidel Castro,* after subjecting his countrymen for a quarter of a century to the burden of an extravagant war economy, may yet realize that the Cuban people are the only real losers of the botched relations with the United States.

From the discussed revolutionary examples, which could easily be complemented by others, like the Islamic Revolution, one main conclusion may be inferred. It is not through revolutionary struggle, or fanned-on class hatred and destruction or costly human sacrifices and martrys that mankind's problems are likely to get solved. Cynics blame the false revolutionary leader, who just uses the revolutionary phraseology and the promise of a better tomorrow as disguised tools for personal aggrandizement, for a macho's shortcut to power and prestige. But this thesis of the morally depraved and villainous revolutionary, who preaches destruction and delights in bringing down the societal pillars of the present in order to build upon them a gorgeous future, does not stand up to a critical trial. Most revolutionary leaders of some stature are profoundly convinced of the righteousness of their cause, and they believe firmly that regardless of sacrifices and temporary setbacks, the final verdict of history will be on their side. Nor are the causes for which they call for fanatical support productions of their fantasy. Hunger, poverty, and the other privations to which large segments of people are subjected in our age of plenty for the few, are real enough. Much closer to truth comes *Oscar Wilde's* sagacious aperçu that, "The people who try to do the most good, are those who do the most harm." Blinded by the just society he proposes to construct, the revolutionary is stoically impervious to the enormous human losses and sufferings he produces in order to attain it. Having accepted *Machiavelli's* golden rule that "the end justifies the means," as *Roger Garaudy* aptly put it, "He is permanently tempted by the exigencies of the liberation struggle to corrupt or to destroy the very liberty he is fighting for."[9] This rationale endows revolutionary processes with a cruel and counterproductive dynamism of their own, in which all mistakes, excesses, and crimes are justified by the supreme goals of the cause and the good intentions of its leaders. Failures and setbacks are attributed to the fiendish class enemy rather than to the revolution's false promises, and to the shortcomings and blunders of its leaders. But romantic ideals and good intentions, when married to political dogmatism and economic dilletantism, are poor advisers and sure guiding poles to bankruptcy and defeat.

Earlier in this chapter, I recalled the opinion of more than one illustrious mind of our time, that something good would come from the critical and angry young generation that swept the continents less than two decades ago and that provided the intellectual spark to many of the revolutionary enterprises, devoted to change mankind's intolerable status quo. There was hope, for a brief moment, that maybe the young, free from their fathers' preposterous bunglings that had touched off the carnage of World War II, and with motivations transcending the narrow boundaries of traditional power-politics, may be able to make a valid contribution for overcoming the crisis syndrome and stalemate, into which the world was pushed by new global chasms and confrontations. Now, as we cross the threshold of the mid-eighties, we may as well bury this delusion. In terms of solid accomplishments, satisfaction of the people's basic needs and true economic advances that would have otherwise been impossible, the contribution of the professional revolutionary idealist has been appallingly sterile, and in many countries, quite counterproductive, entailing—as in Chile—a furious reactionary backlash. The impatient world-improvers, and bomb-throwing fanatics, and trigger-happy guerrillas, who have made violence a holy trade, have turned out to be false prophets, and even worse, amateurs of genuine economic progress and general well-being. The paradise they preach turns out to be a tantalizing mirage. Eventually realities take charge, twisting the course of development into directions which the lofty daydreamer had neither anticipated nor desired.

Again I find the extraterrestrial alternative a worthwhile proposition. With but a few exceptions, political establishments all over the world are handlocked by immobilism and the tight grip of the superpower's bid for global supremacy. The economic elites, while thriving on technological innovation, and still in charge of the world economy's topsy-turvy behavior, cling ideologically and spiritually to a world order, rooted stronger in yesterday's anachronisms than in the necessity to shape a new future. Only the world's foremost scientific and technical estates, to whose prowesses I will turn in the next chapter, keep a mind-boggling revolution going. Only a revolutionary scientific breakthrough with a profound meaning for our ominous organizational stalemate may be able to drive the liberating wedge in our civilization's logjam situation. Contact with a scientifically and socially far superior extraterrestrial intelligence may deal man's complacent soul the earth-shattering jolt for reconsidering what he is and where he may go.

This perspective should be particularly attractive to the young generations, who have followed too long the siren songs of false gods and who are slated to play the leading part in this thrilling scientific adventure. Astronomers are pushing back the infinite frontiers of space at a fabulous rate and are unveiling mysteries of the universe undreamt of only a decade or two ago. It will be the fantastic task of tomorrow's scientific and technological elites to make contact with another civilization's reality and to engage in the mutually fertilizing dialogue, which is bound to produce revolutionary changes in our world. For those who care and cannot wait, revolutionizing this world with the bible of justice in one hand and dynamite in the other, here is an enormously challenging and rewarding alternative.

Chapter 8

Science and Technology: Between Glorification and Damnation

God has given man the gift of thought, the ability to explore all knowledge, providing he uses it for the benefit and relief of the state and society of man.

—Francis Bacon

During the past forty years, no other field of human activity has reaped more triumphant victories and has, at the same time, been the target of more criticism and demonization than science and technology. In the mid-sixties, *Walter Orr Roberts* could still declare, "The age of science has been ushered in and with the sure promise that soon there would be food and plenty for all, education for all, freedom for all." But *Barry Commoners's* caustic retort was, "Modern technology gives us everything . . . but it threatens our survival."[1]

Today, in the wake of the breathtaking advances in such fields as microelectronics, bioengineering, and the conquest of space on one hand, and the accidents in Harrisburg, Bhopal, and Chernobyl, the *Challenger* explosion, and the growing stocks of suicidal weaponry on the other, the fronts in favor and against science and technology have become firespitting battle lines. This rift between advocators and detractors of the scientific and technological revolution is causing much frustration and concern. It throws wrenches in the decision-making processes everywhere, spurring heaven-storming R & D projects on one side and careless neglect on the other. In developed and developing nations alike, cataracts of wishful thinking and unctuous moralizations add up to a situation cast in much confusion.

In World War II, science and technology won the contest for freedom and democracy. "Atomic bombs, proximity fuses, advanced bombsight, radar, robot bombs, psychological warfare, and strategic evaluation," according to *James E. Katz*, "all irrevocably changed the face of warfare."[2] But they did more than that. In its aftermath, they created a landslide of euphoric expectations that science and technology would tackle the problems of peace with equal success. They were to spearhead the recovery of the mangled European nations and win—hailed by Marxist doctrine as "productive forces"—new victories for the construction of socialism and Communism. In underdeveloped countries, hope abounded that with the aid of massive transfers of scientific and technological know-how, the development gap would be bridged in no time. Today's appraisals of what science and technology can do are much more sober. The belief that they might provide the magic panaceas to mankind's vexing problems have been chilled. In spite of the grandiose record of scientific and technical achievements, the world's dramatic global issues loom as baffling as ever.

Responsible for this paradoxical development was not and is not the lack of scientific verve or technological ingenuity. On the contrary, during the past four decades a knowledge explosion has taken place in the industrialized world, dwarfing everything man has accumulated in learning from *Democritus* to *Einstein*. The scientific estate, run by hundreds of thousands of researchers, engineers, doctors and planners, and greased by billions of dollars from governmental and private sources, has come of age. In the U.S. total R. & D. expenditures have skyrocketed from $5.2 billion in 1953 to over $110 billion in 1985. The number of scientists and engineers engaged in R & D activities has more than quintupled from 225,000 to over 1,200,000. In other Western countries, the scientific explosion has been no less breathtaking. The Federal Republic of Germany boosted its R & D budget from $2 billion in 1961 to over $19 billion in 1984. "In order to avoid sinking into bitter mediocrity," *President de Gaulle* pushed hard for France's technical and scientific research, and in 1982, *President Mitterand* launched the ambitious project "to convert France into the third scientific and technological power, behind the U.S. and Japan." Great efforts have also been made by Great Britain, where Prime Minister *Harold Wilson* exhorted his countrymen "to forge Britain's future in the white heat of the technological revolution." But the most spectaculor scientific development took place in Japan. Starting in 1945 practically from scratch, it propped up an R & D capacity, second only to those of the U.S. and Soviet Union today. By 1982, Japan had more industrial robots installed than all other countries combined and was pushing hard to overtake the U.S. in high tech microelectronics and biogenetics. Nor did the Soviet Union stand behind. It is estimated that by 1980, its R & D expenditure already topped that of the U.S., its scientific working force probably doubling that of the U.S. China, too, made spectacular strides in modernizing its science and technology, especially since 1978, when solid measures were taken "to increase rapidly research facilities and the number of research workers until 1985," and "to catch up or surpass advanced world levels in all branches" by the year 2000.

Pressed by strategic and economic exigencies, every year, governments of OECD countries and the COMECON bloc are pumping bulkier resources into priority R & D sectors, such as defense, space exploration, energy, microelectronics, and telecommunications, reaching almost 50 percent of total R & D expenditures in countries like the U.S. Private enterprise chips in the other half. The U.S. automobile industry embarked upon a gigantic plan, involving $50 billion dollars in five years, for robotization of their production plants, and the German and French competitors rapidly followed suit with similar modernization projects. Leading Japanese enterprises as Fugitsu, Hitachi, Sony, and Kawasaki are known to spend up to 15 percent of their gross income for R & D, which helps explain the lead their products have attained in world markets.

A dazzling new world of science-cities is emerging. The "Island of France" near Paris, "Technopolis" in the vicinity of Bari, and "Tecnocity" near Milan in Italy are the new would-be boosters of scientific inventiveness, competing with Silicon Valley in the U.S., and Novosibirsk in the Soviet Union. In Japan, too, a new "Science City," with fabulous research facilities and housing for more than 10,000 scientists and technicians, has been built practically overnight.

The physicist working on nuclear fusion; the electronic engineer designing chips with a million or more micro circuits; the spacecraft expert working on new space exploration or exploitation methods; the physician investigating new ways of the transplantation of organs; and the microbiologist, exploring new potentialities of gene manipulation—they are the real heroes of our age, capturing headlines and awards. Scientists have become the world's undisputed vital elite, a new priesthood. Science is the religion which is changing the world.

This stunning development is only the last stage of the exponential growth of science and technology during the past 250 years. Since *Isaac Newton,* the number of scientists has doubled every twelve years. Of all the scientists that ever lived, four-fifths are alive today. *Don Price* has calculated that "the average growth rate of scientific knowledge per year is 6.5 percent doubling every eleven years." The result was a drastic reduction of the time needed for the development of new products and processes. While it took 112 and 56 years, respectively, to develop photography and telephony, radio was developed in thirty-five years, television in twelve years, and it only took three years to develop the integrated circuit. This phenomenon highlights the striking acceleration of the pace of scientific discoveries and technological innovations that keep us gasping at the gushing stream of new insights in the mysterious workings of nature, the avalanche of new consumer goods, new cost-saving production methods, new therapies for diseases, but also the always more mortifying weapon systems of military R & D establishments.

On one hand, scientific and technological progress has brought humankind incalculable goods and benefits. In the Western welfare societies, it has given man an enviable range of material comfort and a continuously rising standard of living. It has opened up to him the macrocosms of space and the microcosms of the gene, given him access to almost inexhaustble energy sources, and presented to him in microelectronics and biofactories time and labor-saving, as well as ecology-friendly, industries. Science and technology have multiplied man's communication capabilities, making massive instant communication with almost any spot on the globe a reality. They have placed at his disposal the computer, which enhances manyfold his intellectual potentialities and his efficiency in an unprecedented variety of fields. Robots have brought relief from physically dangerous and tedious jobs and are steadily increasing productivity rates.

In the medical field, advances in immmunology and methods of early diagnosis have reduced the death toll of many diseases. Life expectancy rates have risen in almost all countries, and due to the development of a new surgical branch, some fifty ill or badly functioning human organs can now be restored with artificial parts. In the entertainment field, too, television, cable, VCRs, and a whole generation of electronic devices are filling man's leisure time with new meaning and joy.

And all this is only the beginning. Even more fabulous scientific and technological breakthroughs and advances are to come. Controlled nuclear fusion processes will give man access to unlimited cheap energy sources. Gene-splicing processes, already under investigation, will lead to producing nitrogen, synthetizing plant life, to steep productivity increases of livestock, and probably even to the development of new, useful plant and animal species. According to *Philip H. Abelson,* editor of *Science,* we are already "in the beginning phase of exploitation of the ability to engineer proteins."[3] Discoveries of this importance will bring us closer to the goal of eradicating famine and other nutritional deficiencies. Foolproof ex-ante sex determination will probably soon be available to couples wanting babies and, as Prof. *Roy Walford* from UCLA and the French expert *Jean Bourgeois-Pichat* contend, treatment of "DNA" with ultraviolet rays and other techniques—retarding the aging process—is likely to raise human life expectancy up to 150 years.

Even more fantastic advances are foreseen in computing. "In only ten or twenty years perhaps," stated *Steve Sinclair,* president of Sinclair Research Ltd., to the U.S. Congress in 1984, "we will be able to assemble a machine as complex as the human brain. . . . In decades, not centuries, machines of silicon will arise first to rival and then surpass their human progenitors."[4] Robots are bound to run production, services and household chores and will augment man's leisure and learning time manyfold.

But there is also another side to today's frenzied scientific and technological progress. In his epilogue to *Small Is Beautiful,* author *E.E. Schumacher* wrote, "In the excitement over the unfolding of his scientific and technical powers, modern man has built a system of production that ravishes nature, and a type of society that mutilates man." *Herman Kahn* also warned that science and technology "now appear to raise a general threat to the continuation of our civilization," and the OECD expert *E. Cooper* put it bluntly already twenty years ago, saying that "for the mass of humanity, science has probably brought more trouble than gain."[5] Today most criticism is directed at the following five principal target areas:

a. *Military R & D.* In all major industrialized nations, military projects gobble up staggering amounts of R & D budgets. The U.S. budget proposal for 1987 includes $43.3 billion for the development of a frightening arsenal of new nuclear, laser-guided, chemical, biological, and other weapons for suicidal mass destruction, a 20 percent rise over 1986, whereas health-oriented research has experienced a substantial drop.

b. *Destruction of the Ecosystem.* Unchecked application of modern technologies has produced air-polluted cities, rivers and lakes poisoned by industrial wastes, the slow death of forests due to acid rain deposits, and high levels of toxic elements in agricultural products, dangerous not only to many wild life species, but to the health of man himself. Excessive heating of the atmosphere, harm inflicted on the ozone layer in the upper stratosphere, and above all, radioactive contamination, due to faulty methods of waste disposal, leaks, or accidents—as Chernobyl has illustrated—are other hazards and legacies of today's rampaging technological boom.

c. *The Dehumanizing Effect of Informatics.* There is considerable concern that growing dependence on intelligent machines will not only increase unemployment and social inequalities, but that it will also dehumanize man's individual and social life, turning him into a soulless accessory and subordinate of smart automats and depriving him eventually of both, his creative vitality and his freedom.

d. *The Dangers of Biological Engineering*. Not only religious leaders but also groups of natural scientists worry that manipulations of genetic material and experiments with recombinant DNA might lead to dangerous bacteriological developments and escapes, possibly lethal to humanity. The biochemist *Erwin Chargaff* warned that "we should let our fingers go from matters of life and death," and in many advanced countries measures have been introduced to control risky research and experimentation in the novel field.

e. *Development Gap between Nations*. Finally, it is charged that the vertigineous scientific and technical revolution has benefitted principally the industrialized rich nations and that these benefits have not trickled down proportionally to the poor developing countries. Advocates of this thesis even hold that Northern science and technology are the main culprits for the growing development gap.

Other critics, such as the Frankfurt school, have gone to great length arguing that through the false lure of the consumer market the prodigious accomplishments of science and technology have turned man in the West into a spineless slave of the capitalist system, making him feel more and more alienated in a world of sheer technicalities and profit motives. In a different vein, *President Eisenhower* warned in his farewell address about the danger that "public policy could itself become captive to a scientific-technological elite." And there is also the popular, rather biased, undercurrent about the queer lot of scientists, either hopeless idealists or politically unreliable eggheads, pursuing all kinds of wicked goals, as plotting to take over the world—a science fiction favorite.

Science and technology may thus be seen as having too different faces, one holding out to man many treasures and enticing promises, the other holding in store for him not a few disadvantages and threats. This Janus-like character explains why opinions on the role which science and technology are supposed to play today and in the proximate future, are so awfully divided.

On one hand, the optimists proclaim that, existing the needed political will, it is within the grasp of today's physical, biological, medical, and engineering sciences to lick many of the world's most tragic problems from hunger and destitution to underdevelopment. "There are many natural things," admonished an euphoric *Arthur C. Clarke,* "that should be stamped on, hard—and much that is artificial, that should be given the utmost encouragement."[6] Accepting this brutal but probably prophetic line of thought, the possibility of designing bacteria capable of producing hormones or proteins, of combining nitrogen or performing photosynthesis, in order to increase medical supplies or the productivity of agricultural crops, may not seem like such bad ideas. The artificial heart planted by *Dr. Robert Jarvik* in the chest of *Barney Clark* in December 1982 markes the take-off point of another revolutionary strand in surgery. The fully automatized industrial plant is already a reality in Japan, and the car guided along the highway by laser-beams and equipped with automatic brakes, is already on the drawing board. Why, then, should the development of an artificial placenta, relieving women with inborn deficiencies for a potentially dangerous pregnancy, or of artificial limbs and organs for those who need them, or of drugs killing pain or improving memory, be considered such sacrileges? Bio-engineers have already grown a mouse twice its normal size, and in 1984, British scientists created a hybrid creature half-sheep half-goat via gene manipulation. Why, I ask, should it be such an offense to create an ape capable of doing the gardening chores, or a cow twice its present size, and able to give the double amount of milk? *Erwin Chargaff* scorns at the idea "of developing a man who drinks water and urinates oil." And I agree, though on second thought, and considering our depletable energy sources . . . why not? It wouldn't have to be a man. The German biochemist *Friedrich Cramer* scoffs at the foolish popular notion that molecular biology may be trying to produce "a genetic Frankenstein or Dracula, an astronaut without legs or a combination of *Einstein, Mozart,* and *Gregory Peck* in one person."[7] But why not, if genetically feasible, make man immune to cancer or AIDS, or twice his present size and strength and with a life expectancy 250 years or more, or even a man with wings? Blasphemous propositions? Maybe, but I wonder how many of the mortally ill or weak would not consent gladly. And if ever the dream of Icarus came true, why should a flying man be a bigger offense to nature than noisy jets and destructive rockets?

On the other hand, the moralists and skeptics counsel caution and restraint, calling on scientists doing military research to abandon their Faustian deal, and on bio-engineers to stop meddling with the sacred essentials of life on ethical grounds.

In this heated debate, one important aspect is often forgotten. With regard to the rugged impetus of the scientific and technological revolution—do we really have much of a choice? *W.W. Rostow* showed that science follows a "logic of its own," striving for "increasingly more sophisticated tools for observation, measurement, and experiment."[8] It follows its own dynamics, constantly pushing the "frontiers of knowledge" further back, bent on penetrating ever deeper into the codes in which the mysteries of nature are written. The continuous introduction of money and labor-saving new production processes and new products, as *Joseph Schumpeter* already taught in the forties, driven on by the profit motive and competition, is the very motor of growth and progress in the market economies. So dynamic is this process of technical innovation and product substitution that, as sales statistic of U.S. consumer goods indicate, 40 percent of the products have entered the market only during the past five years. Recalling the evolution from physical labor over mechanization to automatization, the German science journalist *Dieter Balkhausen* only states a truism, affirming that "the machine will substitute the man on the job, when it works cheaper and better." This is exactly what the microelectronic revolution is about. Introducing computerized data processing and robot systems at a feverish rate, Japanese automobile makers produced in the mid-seventies twice as many cars as their European competitors and outstripped Detroit. The use of computerized office equipment increases productivity and profitability levels. The industrial robot not only reduces labor costs, but also improves quality and permits a reduction of material and scrap. No enterprise can afford to ignore these advantages. The price for neglect is eventual capitulation to the technologically more aggressive competitor. Man is master and slave of technical progress at the same time. Lecturing to labor leaders on the microelectronic revolution, a German Secretary for Research and Technology alluded to this compulsion, stressing that, "Even if we wanted to, we could not stop technical progress of electronics and electronic data processing. . . . Whoever believes that he can afford to ignore the dimension of international competition, will be overtaken and eliminated from the race." Nations, unable to keep up with today's new round of the scientific and technological revolution face this dilemma. The loss of competitive capabilities and economic vitality, unemployment, and the erosion of living standards are their bleak perspectives. The acute concern of countries like France, Great Britian, and the Federal Republic of Germany that they may be losing out in this race due to the obtained lead by the U.S. and Japan, and the strenuous efforts of the European community to endow its R & D efforts with crash programs like ESPRIT, RACE, and EUREKA more verve and muscle, are clear indicators that they have understood the writing on the wall. "Either we conquer microelectronics," warned a high-level British government official, "or we join the underdeveloped countries at the turn of the century." Furthermore, within the global contest between the U.S. and the Soviet Union, the competition for scientific and technological superiority plays the crucial role, and acts as a powerful stimulus for ever expanding R & D endeavors. For either side, the stakes are too high to take a chance. It is survival and victory or the definite retreat from the stage of history.

There is little we can do to stop or change the course of this self-impelling scientific and technological momentum and its underlying rationale. Advocators of zero-sum growth, predicted *Herman Kahn* in his book *The Next 200 Years,* and those who propose stopping or slowing technological progress won't have a chance. *Balkhausen* rightly compares Silicon Valley with the "braintrust of the revolution."[9] All this underscores that the potent thrust of science and technology, as systems of knowledge inextricably interwoven into the economic and political fabric of our time, which they continuously convulse and shape, cannot be changed at our will as the helm of a boat in calm waters. It changes the world and our lives on a scale and at a pace unprecedented in history, that no political revolutionary can match, and whether we like it or not. Welded to science and technology, our civilization has boarded a train running on a one-way track and there is no turning back.

The scientific researcher, motivated by material rewards and fame, striving hard for a breakthrough; the industrial corporation, forced to crank out new, better, or cheaper products to stay ahead in the cruel competitive race; the military R & D behemoths, engaged in a deadly wrestling match for supremacy and the ultimate weapon; and governments, responsible for the performance of their nations' economies, exports, employment levels, and the well-being of their people—they are all subject to the same coercive forces that propel them forward in the never-ending, spiraling race of scientific and technical progress. It is this coerciveness and reckless dynamism built into the springs of Western capitalism, and which rules its relations with the Communist East, that made *Einstein* complain in 1950, "The man of science has sunk so deep that he accepts the slavery imposed upon him by the national states as an unavoidable fate." Expressing a similar concern, *J.D. Bernal* had warned already in 1929, "Scientists are not masters of the destiny of science—their curiosity and its effects may be stronger than their humanity."[10] The development of the atomic bomb and the constant perfection of today's terror-inspiring nuclear arsenals and other means of mass destruction, give *Bernal's* prescient statement an ominous validity. It is no less true, however, that leading physicists, who were involved in the development of the atomic bomb, such as *Leo Szilard, Eugene Rabinowitz,* and *James Frank,* had pleaded vainly for international control and that more than a few prominent scientists have refused to participate in U.S. military R & D programs. Granted that politicians, not scientists, were responsible for Hiroshima and Nagasaki, but scientists provided the initial information for constructing the bomb, and they built the A-bomb, the H-bomb, and all the other nuclear artifacts since.

Many so-called humanist scientists therefore ride scorching attacks against the alleged golden rule of today's scientific and technological efforts "what can be done, must be done." They hold this supposedly "diabolic maxim" responsible for deadly and harmful R & D activities and try to rouse public opinion and government action against research and experiments, often shrouded in suspect secrecy. secrecy.

But does resistance to this maxim on conscientious grounds provide us with a realistic and viable alternative? *Friedrich Cramer* proposes "to grant research a broad uncontrolled field of action, but to submit subsequently technological application to public control." *Erhard Eppler,* a social democrat and former Minister for Economic collaboration of the Federal Republic of Germany, went even farther, arguing the "governmental policies cannot confide in autonomous technical processes, limiting their response to recording their positive effects and lamenting the negative ones."[11] But this clearly smacks of socialist regimentation of science and technology. Freedom and secrecy of corporate R & D efforts are not only fundamental principles of the market economies. Bureaucratic meddling in free inquiry—as recent experiences in leading Western European countries have amply demonstrated—conspires against its very purpose, stifling scientific and technical ingenuity and muzzling entrepreneurial aggressiveness, both prerequisites sine qua non for scientific and technological excellence and a fine economic performance.

However, there are other reasons that should put us on guard against the pitfalls of anti-science witch-hunting. According to *Arthur C. Clarke,* the great lesson of our age is: "If something is possible in theory, and no fundamental scientific laws oppose its realization—then sooner or later it will be achieved—granted a sufficiently powerful incentive."[12] The dizzy pace of today's microelectronic revolution bears this judgment out. It is pushing the advanced countries relentlessly into the arms of the "information society," regardless of the angry outcries of contemporary Luddites against its impact on employment levels and traditional life-styles. Bio-engineering, too, already heralded as the most dynamic industry of the twenty-first century, is expanding vigorously, as the production of an ever-widening spectrum of products, among them insulin and cancer-fighting interferon via cloned bacteria, shows, in spite of critical attacks on ecological and ethical grounds. In May 1984, *Judge John J. Sirica* halted an experiment of the University of California at Berkeley, to make potato plants more tolerant to frost, by spraying them with a genetically altered bacterium. But in the long run, no one will be able to paralyze research efforts in such promising fields as agricultural genetics, aiming at the development of self-fertilization, more nutritious and more resistant plant life, and of microbiological insecticides and proteins, useful as food for animals and humans. It is sheer daydreaming to think that progress along the blazing trail of microelectronics and bio-engineering, promising even more fantastic breakthroughs in the decades ahead, can be hamstrung by legal blocks or moral jitters.

The late *E.F. Schumacher* once observed, "Wisdom demands a new orientation of science and technology towards the organic, the gentle, the nonviolent, the elegant, and the beautiful." But realistically speaking, what is this wishful proposition to mean? Is the nonviolent path in our conflict-stricken world perhaps an attainable goal? Is the abundant leisure time, which *Schumacher* admired so much in Burma, a worthwhile alternative? In the underdeveloped countries I know, leisure time is mainly due to lack of opportunity and employment, to outright poverty. What people in these countries really want and need is work, income, production, goods to buy and export, all that which symbolizes material progress and which the skeptics of technology abhor as the personification of evil.

Friedrich Cramer also calls for a new asceticism and asks man "to abandon material progress, to renounce on the absurd goods of civilization, on chrome and lac, on super cleanliness, on the silliness of fashions, on overeating and welfare-alcoholism."[13] But he, too, is throwing the child out of the bathtub together with the water. Without all this technical and material progress, the motor of the market economies would have stopped running a long time ago, life would be stagnant and unbearable, creative vitality would have been drained from society, and social unrest and revolution would have been the inevitable consequence. Nor can such demands stir enthusiasm in the poor Third World countries, where clearly more material and technical progress is needed, more production and consumption and not less. *Balkhausen* sees it this way too, stressing "Technical progress and prosperity are not everything—but without them everything is nothing." *Jacques Monod* also acknowledged that "Modern society has to be grateful to science for its wealth, its power and the certainty that man, if he wants to, will have even greater riches and possibilities at his disposal."[14]

It is not less science and technology we need, but more of it and, as *Fritjof Capra* postulates, more geared to the qualitative needs of man and more respectful of our ecological niche.[15] Safer atomic plants, controlled disposal of radioactive and other industrial wastes, reduced exhausts and toxic smoke from combustion engines, and development of insecticides and fertilizers not injurious to man and animal life are justified and viable demands. But trying to extinguish or douse the flame of curiosity, which has permitted man to step out of darkness and to illuminate his existence with ever-widening conscious scientific knowledge, and to make life always richer with his indomitable knack for technical innovation, would be nothing short of a crime.

One decisive shortcoming must be acknowledged, however. Modern science and technology is a monopoly of the industrialized nations. Early hopes that with their aid, Third World countries would be able to leapfrog from the stage of the wooden plow to the industrial age and welfare society have been dashed a long time ago. Though developing countries claim more than two-thirds of the world's population, their share of the world's total R & D expenditures barely reaches 4 percent and only 13 percent of all R & D scientists and engineers are employed in the poor South. National R & D expenditures, averaging 0.1 to 0.5 percent of GNP, compare poorly with the rate between 2 and 3 percent in advanced nations. The real scientific and technological lag is, however, even more distressing, if comparative quality of R & D activities is considered and the fact that, with the exception of the "Four Tiger" countries in the Far East and a few other threshold nations, no research is conducted in developing countries in the strategic "high tech" fields. Not only does the Third World continue to be dependent on technology imports; the majority has callously neglected to develop a proper scientific and technological capacity. Due to the tremendous pace of the scientific and technological revolution in the North, attempts of the lagging South to catch up seem almost hopeless. The cost-reducing and quality-improving microchip and robot are already eroding the "comparative advantages" of the developing countries, as low raw material prices and wages, reducing thereby the competitiveness of their manufactured products on world markets and making it almost impossible for them to service their foreign debt and promote dynamic development processes.

In the seventies, the much hailed "intermediate or appropriate technologies" acquired almost mystic qualities, especially in some industrialized countries. The idea that "small is beautiful," closely associated with *E.F. Schumacher,* and that developing countries should favor simple, labor-intensive, rather than capital-intensive technologies, was merchandized as a genuine panacea of Third World ills, but it has become quieter about it since. Not only have Third World spokesmen voiced suspicion that "intermediate technology" may be another scheme of the wealthy nations to stall industrialization of the poor agricultural countries and to preserve their markets for their own industry; but it is now widely accepted that, though intermediate technology may play a positive role on the local level, it is ill-suited to cope with the basic problems besetting most Third World economies—inefficiency, low productivity rates, and sagging competitiveness. *Jean-Jacques Servan-Schreiber* was probably right, pointing out that the Third World's salvation is not "intermediate technology," but rapid assimilation of informatics and modern telecommunications to boost productivity and exports, but few countries, with the exceptions of Brazil, India, and a few others, are undertaking the necessary efforts with the aim of creating self-sufficient research and know-how potentials in these and other revolutionary scientific fields. And while nationalist chauvinists keep indulging in vociferous but sterile philippics against the bewildering pace of the technological revolution imposed by the R & D trusts of transnational corporations, science and technology ride in the rich North and the poor South on different trains, going in opposite directions.

From this sketchy analysis, we may draw the conclusion that, in spite of their grandiose achievements, neither in the developed nor the underdeveloped world do science and technology serve as harbingers of a more peaceful and safer world order. While increasing man's mastership over the forces of nature manyfold and handing him an ever wider array of amazingly useful and entertaining products and gadgets, they have at the same time reduced our quality of life and survival chances and widened the existing development gap. But the fault lies not intrinsically in science and technology itself. Natural and engineering sciences can only help solving man's pressing social, economic, and political issues to the extent that there is the political will and professional capacity to use them for these worthy ends.

Unfortunately, the record of social sciences has been disappointing. It is said that among social sciences, political science, economy, and philosophy are particularly lagging behind the much more buoyant natural, engineering, and medical sciences. They have produced nothing remotely comparable with atomic energy, the microchip, or a biofactory. They have been able to record our crisis-syndromes, but contribute almost nothing to their continued escalation to the present nuclear stalemate and ghastly North-South gap. "The social sciences," said *Carl F. von Weizsäcker,* "still have to earn their recognition in the republic of scientists."[16] Maybe this is because trying to understand and rationalize the inner workings and functioning of the conscious and intelligent species "man" is a much harder task than probing the functioning of unconscious and unintelligent matter and energy. Generations of economists, sociologists, and political scientists have done an admirable job, diagnosing the fundamental ills of our age and submitting proposals for overcoming them, but they usually lack the political muscle for making these proposals work.

The psychoanalyst *Erich Fromm* thought that his "messianic utopia," of a "humanity free of economic coercion, war, and class struggle, living together in solidarity and peace" could become a reality, if we could only marshall to this end "the same quantity of energy, intelligence, and enthusiasm which we have mobilized for our technical utopias."[17] But pious wishes and propositions such as these are useless as guidelines for pragmatic and viable political action. It is the job of the political, economic, and social futurologist to tell nations and the world which path to take that may lead them to a better and safer future, but even more important is to show them *how* to get there. The portrayal of visionary paradises is not enough; it is necessary to show how their transformation into reality may become not only desirable but a political feasibility.

Excessive optimism with regard to the capacity of science and technology to do away with mankind's fundamental problem-issues, should therefore be put on guard. The hopeful futurologist who believes that microelectronics, bio-engineering, and new materials will revolutionize the world and make it a more just and happy place to live, and who gleefully collect evidence, showing that this process is already well under way, will have to stake his optimistic scenarios in a more remote future. Powerful conservative forces still impede science and technology to place their gigantic potentials entirely at the service of the development of a superior, more rational world order. The following five, I think, are of decisive importance:

a. *The Struggle between the Industrial Giants.* The fierce competition between leading enterprises of the U.S., Japan, and Western Europe, that devours huge R & D outlays for new products and processes, follows a logic and dynamism of its own, leaving no room to pusillanimous moralizations, and will keep subjecting nations and economies to its rationale in the foreseeable future.

b. *The Driving Force of Consumerism.* There is no indication that the urge of consumption, to satisfy real or superfluous needs, fanned on by the distributive system and media, that is the energizing force, without which no productive system could survive and keep its motor under steam, is growing weaker. On the contrary, consumerism is not only increasing in the Western welfare societies, but also in socialist countries.

c. *The North's R & D Monopoly.* The industrialized North has gained such an overwhelming scientific and technological lead, that it remains questionable whether the poor South will ever be able to catch up. Everything indicates, furthermore, that the disproportionate R & D growth rates in both halves of the globe will continue well into the 21st Century, making the division of the world between the scientifically and technologically advanced and the retarded nations even more pronounced.

d. *The Global Confrontation.* As long as the confrontation between the two superpowers rages on, no slackening of the accelerated race in military R & D may be expected. Chances that this confrontation will be shifted in reverse gear, permitting a reassignation of the gigantic military R & D expenditures to constructive ends, must still be considered slim.

e. *The Anti-Science Backlash.* Political reactionaries and religious charlatans, especially in the U.S., are beating the drums against science and the scientific method. From the moral objectors to atomic energy and genetic engineering, the Creationists, followers of *Berlitz* and *von Däniken* and believers in the occult, there is a picturesque alliance, asking the scalp of science and proposing to substitute its rationality with a mixture of emotional ecstasy and blind fanaticism.

The direction and performance of today's science and technology is therefore inextricably intertwined with basic forces and trends, shaping mankind's distraught political, economic, and cultural reality. As powerful as the scientific estate seems to be, it is far from operating in a realm of freedom, but is itself caught in the web of our civilization's fundamental contradictions.

In the near future, it is unlikely that science and technology will recover the girdle of innocence, which they may have lost forever, if they ever possessed it. The scientific estate will not be able to shake off the fetters to which it is welded. It will keep pushing toward the unknown, and along its journey, it will keep showering mankind with wonderful gifts. But unless man's fundamental political framework changes profoundly, it will not be able to give him a more peaceful and just world. Whether fear is the ultimate cause that makes it tick, forcing man to explore existence and to search for explanations, as *Jacques Monod* suggests, or whether our innate thirst for more knowledge just obeys the biological urge to expand our control over the environment and the forces of nature, is an academic and philosophical question. What matters is, that without this driving curiosity to unlock God's secrets and the determination to turn them into useful tools for our material well-being, man would stop to be man. *Arthur C. Clarke's* admonition, "Civilization cannot exist without new frontiers; it needs them both physically and spiritually," rings today with the same force as twenty-five years ago.

Science and technology may therefore yet give us the keys to unlock the door that may lead us out from our present stalemate situation. Discussing the chances of contact with extraterrestrial intelligence, *Carl Sagan* envisions the possibility that we may learn a worthy lesson from the stars. "It is possible," he maintains, "that among the first contents of such a message (from ETI) may be detailed descriptions for avoidance of technological disaster, for a passage through adolescence to maturity."[18] Though I think that man should and will be able to lick his developmental problems by himself, I am convinced that contact with a superior civilization would have a tremendously stimulating impact on world conscience. By helping to revolutionize our basic organizational order, contact may not only furnish science and technology with miraculous shots in the arm, but free it eventually from the subservient role of upholding an anachronistic order, to which it is still condemned today.

Chapter 9

The Escapist Urge

Myths are machines for the suppression of time.
—Claude Lévi-Strauss

Ignorance is the most delightful science in the world, because it is acquired without labor or pains and keeps the mind from melancholy.

—Giordano Bruno

It was *Sir Bertrand Russell* who first pointed out that the victorious march of science and rationality would only induce man to search even more profoundly for ultimate answers in mysticism. More than fifty years ago he predicted that the scientific age would come to an end, and that people would turn away from physics to metaphysics, because the former would no longer be able to instill in them a belief in a meaningful present and hope for the future—something that is most precious for man.

Sir Bertrand Russell was off the mark with regard to the first prediction, but he was dead right in foretelling the amazing boom which the world of pseudo-science, the supernatural and paranormal, the magical and occult and the fantastic and weird enjoys today, even in nations belonging to the scientific avant-garde of our time.

In *Future Shock,* author *Alvin Toffler* called attention to this flabbergasting phenomenon. "We witness," he wrote, "a garish revival of mysticism. Suddenly astrology is the rage, Zen, yoga, seances, and witchcraft become popular pastimes.... Existentialist oracles join Catholic mystics, Jungian psychoanalysts, and Hindu gurus in exalting the mystical and emotional against the scientific and rational."[1]

Since Toffler's bestseller in 1970, this enchantment with the kooky, metaphysical, and emotion-loaded has reached preposterous dimensions, especially in the United States. Suspect theories and fads as resurrection, telepathy, precognition, astrology, psychokinesis, time travel, levitation, biorhythm and primal scream therapy, to mention just a few, hold a fascinating grip on the imagination of millions of people. UFOlogy has become a popular quasireligion. Parapsychology is prospering, especially since it gained membership in the AAAS. The purported mysteries of the Bermuda Triangle keep duping large audiences and Velikovsyite "catastrophism" and crackpot "doomsdayism" are in the vogue, confounding public opinion and scaring the wits of people. The belief that the end of the world is near has today a larger following than *Nostradamus,* the grandfather of today's peddlers of apocalypse, ever dreamt of possessing. That credulous disciples of the Adventists, Jehovah's Witnesses, and of *Brother Emann* have vainly waited in 1844, 1914, and 1925, and again in 1955 and 1960 for the world to come to a God-ordained shattering end, is conveniently forgotten. The popularity of black magic, the widespread faith in the hocus-pocus of fake healers and the belief in evil spirits and ghosts, successfully exploited by *Jay Anson* in his book, *The Amityville Horror,* are other traits of this macabre trend.

Eastern influences such as Transcendental Meditation, Tao, Zen, Yoga, and Swami have also succeeded a penetrating impact on Western societies since the early seventies. Hindu and Korean gurus have established themselves almost as gods, to whose imaginary paradise the younger generations in particular keep peregrinating in search of deeper meaning and spiritual bliss. Retrogressive currents of fundamentalism, too, flaunting everything true science stands for, are celebrating an astonishing revival, above all in the United States. Gallup polls, conducted in 1980–81 showed that 53 percent of Americans believe hell exists, about two-thirds are convinced that there is life beyond death, and 71 percent think that "there is a Heaven where people who have led good lives are eternally rewarded."[2]

If this fascination with the mystical, magical, and pseudoscientific were just a random phenomenon, affecting only a marginal segment of the population, it could probably be shrugged off with a faint smile. But the quackery of the witchcraft practitioners, astrologists, soul-saving gurus, fake prophets, fortune-tellers, and other canvassers of the paranormal and metaphysical, preying like buzzards on the most intimate cords of the human psyche, transcends the realm of the strictly personal and private and is an issue deserving public concern. It not only fosters retreat from haunting reality, opening the door to hallways of illusionary panaceas and paradises, but it represents a massive attack on the very essence of the scientific and rational outlook that shaped the Promethean dimensions of our modern world, and a strong shot in the arm of pro status quo forces, attitudes, and behavior patterns. However bright and sacred these hallways may beacon, to seek in them the Promised Land of peace and happiness is a dangerous delusion. Instead of helping to attack the real roots of our supercrisis, mysticism only plunges our world deeper into a mélange of exotic obscurantism and chaos. But before venturing a few concluding remarks about this gaudy "myth and cult world," where the wish to believe rather than to know, and the desire to experience emotions rather than the sense of responsibility to tackle collective problems rationally carry the upper hand, let us take a brief look at some of this sprouting outgrowth of irrationality.

Astrology is today more than a fad and a profession; it is a religion. In the U.S. alone there are some twenty thousand astrologers, compared with two thousand astronomers, and horoscopes are carried by the majority of daily newspapers. Gallup poll results indicate that in the late seventies, astrology had between 30 and 40 million believers nationwide.

The claim of astrologers that zodiac signs predetermine human character traits, that they influence their daily financial, romantic, and other enterprises, and that they can predict the future, is unproven and scientifically unsustainable. The imaginary zodiac belt and far away constellations can have no bearing whatsoever on our personality. A test conducted at the Northern Kentucky University in response to a sensationalist article of the *National Enquirer* in January 1980 revealed the flimsy nature of the alleged zodiac sign—personality connection. In a well-documented series, the German magazine *Der Spiegel* disposed of *Nostradamus's* supposed predictions of the two world wars, Hitler, and other predicaments, as mere hogwash.[3] For 1982, French astrologists prognosticated the assassination of the pope, another attempt on *President Reagan's* life, and disastrous earthquakes in the U.S., China, and Japan—none of which occurred—while failing to predict the death of *Brezhnev* and the Falkland Island War. Nor have astrologers been any more successful foretelling the earthquake in Mexico, the eruption of the volcano El Ruiz in Columbia in 1985, or the disaster of the *Challenger* early in 1986. Clearly astrology is a fake science, feasting on the credulousness of people. Regarding its continued success, 186 top scientists, among them sixteen Nobel prize laureates, gave this explanation in 1975: "In these insecure times, many people want to believe in a fate predetermined by astral forces, upon which they themselves have no influence." In tow of the astrology myth, they prefer to receive guidance for their decisions and actions from mysterious forces, rather than from knowledge and rational analysis.

Parapsychology. Some people, its proponents claim, are endowed with extra strong powers of extrasensory perception (ESP), enabling them to read other people's mind, to perform telepathy, or move material objects with their minds. Truckloads of so-called conclusive evidence has been presented in favor of ESP, but—as the American philosopher *Paul Kurtz* has pointed out—so far all attempted repeats of the experiments under tight scientific procedures have failed. *Uri Geller's* stunning performances in bending spoons, supposedly proving his paranormal capacities, have been exposed by *Martin Gardner* and *James Randi* as audience deceptions. Other famous psychics, as *Jean Pierre Girag* and *Judy Knowles,* were inexplicably abandoned by their professed paranormal powers when under strict observation. *Mrs. Kalugina,* a known Russian psychic, has been exposed as a charlatan and convicted for cheating.[4] Parapsychology is another attempt to disparage what science has so far ascertained about man's congnitive processes, and to induce people to rely on their untapped mysterious paranormal forces to perform extraordinary feats.

The Bermuda Triangle Mystery. Few myths have been as successful as the yarn that the disappearance of ships, boats, and airplanes in the triangle between the Bermudas, Miami, and Puerto Rico is due to the existence of a mysterious evil force. According to *Charles Berlitz's* bestseller, it is all the fiendish work of an underground pyramid off the coast of Florida. Not one bit of hard evidence backs up the preposterous claim. As *Larry Kusche* showed, *Berlitz* and other authors consistently twisted evidence of misshaped planes and other accidents to suit their aims, eagerly hyping eerie suppositions, when quite normal explanations were available.[5] The much hypothesized disappearance of Flight 19 in 1945 is now attributed to navigational error. The strange and much exploited plunge of National Airlines' Flight 120 from Miami to Newark on Jan. 27, 1978, from thirty-three thousand to twenty-five thousand feet, when engines suddenly restarted, lost all its mystery when the flight engineer later admitted, that he had forgotten to turn on the fuel boost pumps. *Kusche* offered *Berlitz* a $10,000 bonus if he furnished proof of the alleged pyramid. But his offer was turned down, quite evidently, as the *Skeptical Inquirer* concluded, because the author was unable "to bear the burden of proof to show what he says is correct."[6] This incident shows what the gay concoctions about the Bermuda Triangle are really worth.

Witchcraft Healing. Crank doctors and medicine men perform rituals and phony healing acts by the thousands daily in almost every country of the world. My wife witnessed such a performance in Havana, where a *curandera,* using a pigeon, drove the evil spirit out of the body of a young woman, who had remained childless after eight years of marriage. Using much blood and incense, and invoking her saint, she tricked everyone into believing that the woman's evil spirit had passed into the pigeon, which took it with it, flying away through a window.

How crank doctors really work was revealed by the German science writer *Professor von Difurth,* who filmed the performance of a famous Phillipine spiritual healer, discovering that the supposed wonder healer was applying ordinary pickpocket and magician tricks. Fingers, supposedly penetrating the body of the sick person without leaving a trace, were simply bent at the joints. Coloring substances, genuine blood and screws and sick tissue—purportedly removed during the operation—were administered stealthily by quick-moving assistants.[7] The racket with the sick pays star healers and barefoot quack doctors handsome dividends. The infinite hope and gullibility of the patients are no match for their crafty rituals.

Creationism. Creationists hold that evolutionary Darwinism is a heretical fake and that the biblical version of genesis is the only correct one. Their most grotesque claim is that Earth is not older than six thousand to ten thousand years. This thesis stands in crass contradiction with all scientific findings. Though scientists disagree on minor details, Darwinist evolution is one of science's few undisputed and most honored theories. With the aid of at least three independent scientific methods, the age of our solar system has been established at roughly 4.5 billion years. The dinosaurs, as fossil finds show, died out no less than sixty million years ago. It is estimated that it took Australopithecus four to five million years to become Homo sapiens. The fact that the genetic code is the same for all living species means that they must have originated from one primordial cell and developed over a very long time, stretching some 3.5 billion years. But creationists reject these findings disdainfully. They prefer to believe, as *John Skow* stressed, that the devil put "fossils into rock to tempt man to doubt the Bible," and that two million animal species made it simultaneously to the Near East from such far away places as Australia and North and South America to reach Noah's Ark and that they managed to stay alive on it for a year.[8]

In spite of clinging to such unsustainable explanations, creationism is blazing along a surprising bonanza trail in the United States. In 1981, public schools in Arkansas were ordered to give balanced treatment to creation science and evolution science.

The Sect and Cult Boom. Since the early seventies, an amazing tide of eccentric religiosity has swept the American scene. Unconventional sects and cults with odd rules and habits are mushrooming among the younger, disenchanted generations not only in the U.S., but also in Western Europe. In 1978, *W.S. Bainbridge* and *R. Stark* located in the U.S. alone, 501 headquarters, 167 of these in California. *Alvin*

Toffler thinks the correct figure is closer to one thousand, with a combined membership of three million. The real following is probably much larger.

Among the most influential sects and cults, we should mention the Children of God, the Church of Scientology, Transcendental Meditation, Process, New Thought, the Church of God, Silva Mind Control, the Moral Majority, and the anti-red Christian Crusade. A distinguished place is occupied also by *Oral Roberts's* Healing Waters, the Thiosophists, Spiritualists, and psychedelic groups, and by the movements with strong Eastern traits, as Zen Buddhism and Lamaism, Hare Krishna, the Divine Light Mission and *Reverend Moon's* Unification Church. Since it would be a vain endeavor to present here the plethora of fantastic claims, strange rituals, and spiritual jiu-jit-sus, with which these sects and cults snare their followings, a few sketchy remarks must suffice. As to the questionable merits of training technique such as hopping around for hours in the vain attempt to levitate or to exercise mind control, or of walking barefoot across a path of burning charcoal for experiencing the sensation of success, they hardly need substantiation. Nor is it clear why the costly courses for talent development of the Church of Scientology should deserve credibility, based as they are on such deep thoughts of its leader *Lafayette R. Hubbard* as "time is the primary cause of untruth" and that "on the 9th of May in 1963 at ten o'clock and half a minute, he had visited heaven for 43,891,832,611,117 years, 344 days, 10 hours, 20 minutes and 40 seconds."[9] And what are we to think of *David Berg,* the head of the Children of God, who has tried to discourage teenagers from attending public school at a time when more and better education is a crucial issue for the U.S. to stand its ground. *Jim Jones's* statements that "he was the reincarnation of Christ, Buddha, Marx, and Lenin" and his spurious claims to have resurrected forty members of his sect from the dead, revealed the perturbed mind of this prophet and might have served as a foreboding for the dreadful end that awaited his followers in Jonestown.

The urge to escape from the antinomies and insipidities of modern life is most pronounced in the sects practicing oriental-type mysticism. The head-shaven devotees of Krishna Consciousness, clad in long orange robes, asking kindly for a small donation, are innocent-looking enough. Just how "love for Krishna," to be shown by chanting "hare Krishna" sixteen times daily for each of 108 prayer bends on a string, will drive out "karma" (spiritual demerit), improve the world, and give humankind ultimate happiness, remains a mystery. Nor have the meditational practices and "satsang," the lectures and spiritual discourses of guru *Majaraj Ji* of the Divine Light Mission, ushered in the prognosticated millenium in 1973. His teachings, that the external world (maya) is an illusion and that only his lectures endow the believer with divine light, are unlikely to alleviate our world's ghastly political and economic problems, which are very real and pressing. *Bhagwan Shree Rajnesh,* another succesful guru, asked of his disciples complete surrender of their ego, reasoning, and rational judgement, and utter rejection of the deadly views of *Darwin, Freud,* and *Marx.* They were not to waste their time on such earthly trivialities as biological evolution, sex, and social justice.

In 1986, *Bhagwan*'s empire, which had expanded into more than twenty countries—Rajneshpuram in Oregon had been his latest Mecca in the U.S.—fell apart, among other reasons, because of criminal charges initiated by U.S. judges. The "God-like" *Bhagwan* himself was jailed, but later permitted to return to India, where he promised to create—what else could it be—a new sect!

The core of Eastern mysticism, as in Zen Buddhism, is that the phenomenal world, which we perceive through experience and our mental activity, is unreal and that, as *Amaury de Riencourt* has pointed out, only by overcoming mental activity altogether man can eventually merge, "identify with, and be the Ultimate Object."[10] The pupil of Zen, *Eugen Herrigel* explains, "does not look . . . for any rational solution, having learned that thinking is totally useless and must be eliminated."[11] The objective is to experience "satori," a state of rapture and ecstasy, in which illuminating insights are supposedly reached about the very truth of things, a kind of mental orgasm, in which nothing but this orgasm is real and matters. The convert retreats further and further in a self-centered visionary world, develops a neutral and almost disdainful attitude toward reality, abandoning himself almost entirely to mystical experiences. Those who revel in this "oceanic feeling," says *Riencourt,* " . . . are often irresistibly tempted to abandon the world altogether and plunge forever into this supernal 'beyond,' leaving the rest of mankind to its blind fate."

The UFO and von Däniken Myth. The fanciful creed that the Unidentified Flying Objects reported to have been seen all over the globe are visiting spaceships from advanced civilizations is another popular and widespread religion, but it is in the U.S. where it has reached the highest pitch. The reported sightings already top 60,000. A fetid pro-UFO literature swamps the country. Motion pictures like *Close Encounters of the Third Kind* and *E.T.* fed on the yarn. Even the U.S. Air Force was forced to investigate the UFO craze. Though no evidence for the extraterrestrial origin of UFOs was found, a Gallup poll of 1973 showed that 11 percent of American adults thought they had seen a

UFO. A survey of University of Washington undergraduates, conducted as recently as 1979, found that up to 67 percent of the interviewed agreed with the statement, "UFOs are probably real spaceships from other worlds."[12] The Condon Report, contracted by the U.S. Air Force in 1966 and released early in 1969, identified the majority of the selected UFO reports as "natural phenomena." The few remaining reports, classified as "unidentifiable," have subsequently been explained satisfactorily by independent researchers, who were able to detect their "prosaic" origin. More recent supposed UFO sightings, as the alleged attack of a UFO fleet on four Brazilian jets on May 21, 1986, and the purported interception of a Soviet passenger plane by a UFO in January 1985, quickly lost their glamour, when convincing explanations related to a dying satellite and military experiments became available.

From this negative record of more than thirty years, just one conclusion imposes itself. UFOs are nothing but sondes, satellites, atmospheric phenomena, optical illusions, and concoctions of sensation-hungry or commercially motivated minds. Plain logic tells us that if any of the UFOs had been true messengers of space, their crews—of natural or artificial origin—would have attempted contact and to establish a meaningful dialogue with us. The late *Dr. Allen Hynek,* founder of the Center for UFO Studies in the U.S. and a long-standing prominent UFOlogist, began to backpedal during his last years, wondering why the supposedly highly intelligent aliens never made a serious attempt to make an official contact, which undoubtedly would be rewarding for both parties. Another early UFO enthusiast, the physicist *Jacques Vallee,* also seems to have been assaulted by second thoughts, alleging that after all, "UFOs may not be from outer space," that "they may, in fact, be terrestrial-based manipulating devices," a no less fantasy-straining elucubration.

It is also reasonable to suppose that if UFOs, were extraterrestrial craft and had landed on Earth or penetrated Earth space as often as it is alleged, at least one of the millions of UFO fans would have been able to cash in on the *National Enquirer's* as well as Cutty Sark's offers of a million dollar and a million pound endowment, respectively, to anyone capable of proving that Earth has been visited by an extraterrestrial spacecraft.[13] Oddly enough, this has not been the case. So far, no one has been able to submit a proof, satisfying the rather stiff requirements, namely acceptance by the U.S. National Academy of Sciences in one case, and the Royal Academy of Science of the United Kingdom in the other. Surely, this is not evidence bolstering the UFO hoax.

The bold proposition that Earth has been visited by intelligent extraterrestrial beings in its historic past, a thesis skillfully exploited by *Erich von Däniken* and many of his epigones and eagerly accepted by millions of their uncritical readers, does hardly fare better. Though *von Däniken* has done a painstaking job, compiling a remarkable amount of references and descriptions from ancient scripts and legends, Sumerian and Indian mythologies, and Mayan drawings, none of it stands up to a thorough scientific examination and proves his basic contention. Presenting his case with lots of charm and imagination, he has nevertheless been incapable of presenting a single piece of evidence, a manuscript, a photograph, an unknown but artificial alloy, a superior mathematical formula, or anything that would prove beyond the shadow of a doubt that, in fact, Earth has been visited by intelligent extraterrestrials in the past. Consequently, as long as this conclusive evidence is missing, his luxurious claims deserve a healthy dose of skepticism.

In his latest book, *Have I Been Mistaken?,** von Däniken* admits that some of his earlier claims had been erroneous. But his new claims are even more extravagant. From the fact that the *Mahabharata,* one of the big Hindu epics, contains a lusty description of flying machines and devastating weapons and wars, he quickly infers that spaceships had existed before and that apocalyptic nuclear wars—resembling Star Wars—supposedly with the aid of extraterrestrial technical know-how, had destroyed earlier human civilizations.[14] My readers may judge for themselves whether this latest concoction of van Däniken and a Hindu Sanskrit specialist—without providing a single relic as proof—deserves more credibility than his earlier allegations.

Still, "von Dänikenism" and the UFO myth keep enjoying fanatical followings all around the globe.

After this brief excursion into the jungle-world of the pseudo-scientific, occult and mystical, a stabbing question needs to be addressed: What does this boom of irrationality mean?

In the first place, I think, it would be false to ignore or to belittle it. The statement of the professor of divinity, *Harvey Cox,* of Harvard University that "the cult menace is largely a fabrication or an illusion" can hardly be taken at its face value. Such developments in the U.S. as the admission of parapsychology to the AAAS, the attempted legalization of Creationism in official school systems and the attacks of fundamentalist radicals against Humanism, calling it "satanic in nature," because this movement is committed to science and holds up liberal American traditions, mark a tendency of great concern. Reli-

*Translated from the German edition *Habe ich mich geirrrt?*

gious fanaticism has already led to Nazi-style book-burnings. At the Omaha Christian School, innocent students watched "as their principal read from the Bible as he fed the flames with copies of *National Geographic,* a magazine which he claimed hindered Christian life."[15] The macabre massacre-suicide in Jonestown and the bankruptcy of *Bhagwan Shree Rajnesh's* spurious spiritual paradise are other climactic demonstrations of the self-destructive end awaiting the unfortunate caught in the treacherous web of credulity and the escapist urge.

The backslide into the magic, paranormal, and mystic is the symptom of a deeper malaise, rooted largely in the disenchantment with the overall performance of our Western societies and culture. To the extent that religious fundamentalism, Islamic radicalism, and eccentric pseudo-scientific trends blossom in the atheist Soviet Union, it also underlines communism's failure to satisfy man's basic spiritual needs. Basically this backslide represents a frontal assault on the scientific and rational view of life and the cosmos, to which mankind owes so many of its material, technical, and social advances. The avalanche of the irrational can therefore not be taken lightly. It menaces to corrode the vital core of these societies, undermining the basic motivations, from which its inexhaustable fountains of Faustian energies and inventiveness spring. Basically it helps to produce the following five negative effects:

Retreat from Reality. By accepting the supernatural and magic as the supreme maxim, the believer not only relativizes man's efforts and achievements but degrades himself also to a meek and dependent tool of otherwordly forces and designs. Thereby he acquiesces in the retreat from existing realities. His infatuation with mysterious powers, and the zodiac signs, the coming of a new savior or the little green men from Mars only weakens his capacity to come to grips with the real problems he himself, his community and the world confront, making him adopt attitudes of aloofness and escapism; similar to the way frustration and alienation drive millions of people in the soporific arms of alcohol, drugs, and the TV trap.

Rejection of the Scientific Rationale. By embracing the metaphysical and transcendental as final causes, but also as guiding stars for man's existence and role on Earth, the faithful follower of the sect and cult subculture is not only inclined to interpret past and future in teleological terms of predetermination and fatality, but denigrates and questions man's capacity to make positive use of logic, reason, and intelligence for furthering alternatives of survival and coexistence worth living for. He holds more of the intervention of dark and mysterious forces than of objective knowledge and the application of the scientific method of trial and error and experimentation as an approximation to truth but also to an always richer, better, and juster way of life.

Spiritual Egocentrism. A marked tendency toward a stultifying egocentrism, called by the philosopher *Delos B. McKown* from Auburn University, "methodological solipsism" is another main feature of the sprouting labyrinth of irrationality. Either by negating the validity of scientific findings and outlooks or—as in Hindu Buddhism—by denying reality altogether, the "self" and the criteria "what is true to me is true" are automatically placed in the center of queries, permitting a voluptuous subjective masturbation. Intense brainwashing, practiced heavily by spiritualist movements, propitiate a spectrum of psychic states from solipsist contemplativeness, repetitious chantings, trying not to think at all or striving for pure forms of mental ecstasy, that hypostatize the "ego" and disparage collective values and ideas existing outside the restricted area of the creed's sacred canons. In the same way, quack sciences such as astrology, fortune-telling and transcendental meditation concentrate on the individual's psychic make-up, making him believe in the existence of cloudy potentials for success, health, and happiness in mysterious regions of his soul rather than in the acquisition of collective knowledge and experience.

Experimenting Emotions. Rather than find satisfaction in understanding and scientific insight, the myth and cult conscript wants to experience truth, deliverance, or spiritual bliss in a strictly emotional way. Evidence of this is the high emotional input of participants in seances and spiritual healings, the spellbound faithful atop a mountain, waiting for Armageddon or a UFO landing, but above all, the freakish states of rapture of disciples in presence of their guru or trying to experience "satori." Into this plays also the longing of the psychologically depressed "renouncer" or "relinquisher," the German "Aussteiger," to lead a carefree, uncomplicated life, free from social obligations and constraints of the establishment, and the wanton desire of the hippie-bohemian gone cultist to derive gratifying sensations by delving into emotional nirvanas.

Doomsday Resignation. The preachings of Final Judgments, Earth catastrophes, cataclysmic wars, and the proximate end of the world tends to breed in the faithful believer not only a completely false understanding of our world's realities but also a quite negative and fatalistic attitude toward man's capability to change destiny and to create for the advancement of his species with the aid of his own intellect and will a world befitting to his genius. Beliefs in the prediction of *Nostradamus,* that the world will come to an end in 1999; in *David Berg's* false prophesy, in 1969, that Los Angeles would soon be struck by a disastrous earthquake; and that Comet Cohoutek would destroy the sinful, which prompted hundreds of his followers to seek exodus in Europe; and in the Children of God's prognosis of a final battle between the forces of God and the allies of Satan, only foster the convert's abject surrender to the inevitable.

From this brief analysis of some of the basic character traits of the ongoing myth and cult boom, one rather self-evident conclusion may be drawn. Can the claim of the foremost leaders of this ghoulish trend that they possess the true key to man's individual and social problems be taken for truth? It is from these commercially shrewdly managed, oily and anti-wordly quarters, we may expect pragmatic solutions to the many complex and pressing issues confronting mankind? The answer is evidently no. Except for esoteric recipes for individual escapism and enigmatic promises of metaphysical bliss, the pompous high priests of what *R.D. Clements* has called "spiritual masturbation" have little to offer. Horoscopes, levitation courses, and "satori trances" will not bring us a step closer to a healthier and juster societal order neither in the "have" nor in the "have not" nations. Such beliefs, as in the end of all evil in God's Kingdom, heralded by *Reverend Moon* or in the gospel of Krishna, that by overcoming "karma" and teaching love, we will improve the world and bring happiness to everyone, just remind me of the funny tale of the "Schlaraffenland," intended for children, where fried chicken fly through the air.

The fortune-teller, the prophet of Armageddon, and the preacher of spiritual ecstasy and redemption only peddle escapist "ersatz." The refuge in mystic realms, trust in the paranormal, retreat into soul gymnastics, worship of zodiac signs, and heaving dirt on science cannot be the answer to the world's throbbing wounds and anxieties. However comforting a horoscope and the stroke of a faith healer may be; however intense the feeling of spiritual love and mystical experiences a guru may evoke in a disciple's heart may be; and however at peace with himself and the world the convert may feel, dousing in yogalike self-contemplation, the needed remedies to the survival problems of our species will have to come from our intellect and creative impulses, and not from sensuous hallucinations and spiritual orgies. No matter how deep an individual's identification with the incomprehensible designs of the Ultimate, his belief in ESP, and a sect's harsh commandments, or how overpowering the rapture of his internal chords may be, the cures to mankind's collective dilemmas will have to be of a sturdier material.

It has been said that the sale of spiritual opium is a better business than the sale of opium. The fact that among the leading salesman in this trade there are but few whose modest life-styles and fortunes belie this accusation, should give food to thought.

Mavericks within the scientific community, such as *David Bohm, Fritjof Capra,* and *Renée Weber* hold that the scientific and mystical world views are converging. The American astronomer *Robert Jastrow* asserted that "the essential elements in the astronomical and biblical accounts of Genesis are the same."[16] But when *Bohm* admits that he has been convinced by the Hindu sage *Krishnamurti* that "thought currupts reality" and when *Capra* confesses having been highly impressed by the same sage's insistence that "in order to know truth, one must stop thinking," I wonder whether the opinions on convergence are not built on shallow sand. Science and technology cannot be humanized by sacrificing "rational thought," which distinguishes man from the animal, and throwing it to the hounds of irrationality. The Hindu sages *Aurobindo* and *Gori-Krishna,* sensing that pure calisthenics in mysticism and passive meditation will not resolve this world's problems, advocate a more balance point of view, by giving historic and biological evolution and science a prominent place in their philosophies.

Religious beliefs in the U.S. and in Western Europe are said to have declined during the past decades. But as an opinion poll, conducted by *George Gallup, Jr.,* as recently as 1981, showed, 71 percent of Americans still believe in life after death, and 71 percent and 53 percent, respectively, believe in the existence of Heaven and hell.[17] Answers to a questionnaire posed in 1979 to 1,439 undergraduates of the University of Washington in Seattle revealed that an average of over 65 percent agreed that Yoga, Zen, and Transcendental Meditation were probably of considerable value and that UFOs were probably real spaceships from other worlds.[18] A significant percentage also believed in the value of occult practices as seances and psychic healing, ESP, and astrology. The mentioned backlash of traditional values, fanned on by missionary evangelists, all dazzling TV performers, though trite in content, probably accounts for these high ratings. Chances are therefore high—tossing traditional religious beliefs, the credos of the new sects and cults, as well as other pseudoscientific myths and prejudice all into one hat—that organized mawkish metaphysics and the scurrile penchant for magic and the occult will continue their gay honeymoon well into the 21st century. *Max Weber,* who more than seventy-five years ago wrote that "with the progress of science and technology, man has stopped believing in magic powers, in spirits and demons," and that "he has lost his sense of prophesy and, above all, his sense of the sacred," couldn't have been more wrong.

All this bodes ill for the Western world's capacity to cope with the challenges that lie ahead. *Isaac Asimov's* admonition regarding Creationism applies to the whole spectrum of the myth and cult threat. "With it in the saddle," he warns, "American science will wither. We will raise a generation of ignoramuses, ill-equipped to run the industry of tomorrow, much less to generate the new advance of the days after tomorrow. We will inevitably recede into the backwater of civilization, and those nations that remain open to scientific thought will take over the leadership of the world and the cutting edge of human advancement."[19]

Japan has already taken the technological lead in fields where Americans and Europeans had been the unchallenged number one not long ago. As a special report of *Science 80* indicated, the Soviet Union's science and math curricula and its output of first-rate mathematicians are vastly superior to those of any other nation, including the U.S. The report, "A Nation at Risk," presented by Secretary of Education *T.H. Bell* in April 1983, dealt a shattering blow to American complacency. It concluded that "our once unchallenged preeminence in commerce, industry, science and technological innovation is being overtaken by competitors throughout the world."[20] Other educational researchers such as *Paul Hurd* and *Paul Copperman* arrived at similar blunt appraisals. This appalling drop of productivity of the American educational system cannot be explained away only with organizational and curricula shortcomings. The deeper causes probably have their roots in the corroding effect of the myth and cult offensive against science and rationality and in the coy nonchalance with which the *Reagan Administration* in coddling its assault on the very values that contributed most to America's greatness.

Prospects for a reversal of this general trend are not very bright. The pervasive attack that the domain of the occult, paranormal, and mystical is launching against the rational pillars of our societies is deeply entrenched in the global crisis that engulfs mankind. It feeds upon the atmosphere of fear, insecurity, violence and frustrations spawned by it. Traditional religions and secular ideologies, in the capitalist and Communist world, fail to provide reassuring answers and stimulating visions of the future to the frightened, confused, and spiritually and emotionally vulnerable segments of our hustling but too status-quo-oriented societal machines. Unable to satisfy man's basic needs, their appeal and mobilizing strength has begun to fade, forcing the new generations to turn to alternative panaceas, offered by the pseudo-world of myth and illusion, of drugs and—as *Bhagwan Shree Rajnesh* said—of suicide as the "final orgasm."

As long as the stalemate will keep dragging mankind deeper in its bottomless gorge, it will continue to furnish fertile ground for the myth and cult subculture. The urge to break away from its insane rationales and to escape from its murderous journey into uncertainty will keep serving as a fertile soil. The escapist urge is the symptom of a global malaise, a danger signal that our disarrayed social and world order is impacting profoundly upon the psychic make-up of man at the eve of the twenty-first century.

To counteract the mudstream of the magic and irrational, science, technology, and education must be imbued with a thoroughly humanist message. I think that by promoting the search for extraterrestrial intelligence we can further this great objective. Contact with a technologically and organizationally more advanced, intelligent species would also show us that the creation of superior global orders is neither subject to the fluke of mysterious forces nor the result of resignation in a preordained fate, but that they are the consequence of perseverance and scientific and rational prowess, which intelligent beings have at their command.

Chapter 10

The Global Deadlock

. . . there has never been any conceivable basis for a one-time, general settlement of the military-political stalemate that World War II produced.

—George F. Kennan

Within the cluster of worldwide problems, each, the crisis of politics; the nuclear nightmare; the development gap; the sterile tide of revolution; the Janus face of science; and the urge of people to escape into dreamworlds, is but one facet of the *problematique humaine*, which, in the words of the Club of Rome, confronts mankind. But this portrayal would not be complete if I would not focus attention to the fact that fundamentally our many-checkered crisis boils down to one basic contradiction, overshadowing all others. There are two political and economic systems competing with each other on the world scale. They may be indentified with the United States on one side and the Soviet Union on the other, though the conflict transcends the territorial boundaries of the two superpowers. The global contest is between Western liberal democracy and capitalism, and communism with its centralized economic system. Shades exist on both sides. European Social Democrats strive toward a third model, somewhere in between, the liberal social welfare state. There are attempts, as in China, to modernize communism and give it a more human face. In the Third World, too, a search is on for alternative development styles and ideological guidelines. But this fuzzy groping for plurality at the fringes of our mainstream cannot conceal the overbearing character of the confrontation between the two supersystems.

As organizational orders of human affairs and progress, each of the two supersystems is held by its champions as superior and preferable to the other. Each, liberal democracy and capitalism, with roots reaching back into the eighteenth century; and communism, a product of a much younger vintage, which has arisen as a powerful challenge to the elder order, lay claim to superior excellence in satisfying man's basic material and spiritual needs. Each holds out to mankind luring promises and the assurance that it is in the possession of the best political, economic, social, and cultural remedies for solving our civilization's crisis and guaranteeing peace and prosperity to all. But where is truth? Which of the two is right, by professing to be better suited for coping with the exigencies of today and tomorrow? Which will receive the nod of history? Never have there been questions that were asked more often or for a better reason.

With regard to the outcome of this historical confrontation, the thing that man probably has least is choice. Besides, neither of the competing systems is immutable to change. Nevertheless, I think that, considering the urgent need of a new organizational order for our world, a critical look at the light and dark spots of both systems can reveal interesting insights.

Liberal Democracy. Though Western political democracy as a bulwark against oppression still exerts strong appeal, especially among people subjected to the brutal arbitrariness of despotic and totalitarian regimes, its claim of safeguarding individual freedom and of providing the best political framework for harmonizing individual and common well-being is questioned today.

On one hand, confidence has been slipping in the capacity of democracy to tackle successfully the grave political, economic, and social issues staring nations and the world in the face. In advanced Western countries, governments and traditional party politics have failed conspicuously to address pressing ecological, pacifist, feminist, and other social needs, driving millions of the unsatisfied into the arms of highly critical and often radical "citizen movements." They seem no longer capable, as *Alvin Toffler* correctly charges, to handle the "decision load," thrust upon them by today's societal and world crisis, and so involved in "crisis-management" and "busy tackling the problems our fathers have left us" as the German Social Democrat *Erhard Eppler* impugns, "that they unwittingly keep passing on the solutions for today's problems to our children."[1]

Closely associated with inept government is the general downward trend of government morals in leading democracies. *Arnold Brecht* touched upon this vital aspect, stating "democracy in the Western sense of the term, cannot survive without a high commitment to the common well-being, high moral standards, and maintaining voluntary participation in assuming responsibility . . ."[2] That there are less and less of these qualities to be found in the leading political echelons of democracies is no longer a secret. Watergate and dozens of high level corruption scandals, involving congressmen, members of parliaments, and even a regent and a prime minister in some of the so-called leading

Western democracies, have pulled the wool from the eyes from the last credulous believer in the innocence of our representative system. This way, the basic principle of representative democracy has been dragged into disrepute. Hunger of power, greed, and subservience to special interest lobbies have eroded the credibility of honest government; faith in the electorates has been drained that the elected will earnestly defend the mandate and interests of their constituencies. To this we may add filibustering and other racketeering tactics, which in many alleged democracies are not only exposing legislative bodies to public ridicule and undermining the independence of the judiciary branch, but also effectively stymieing government action.

But an even more fundamental weakness seems to be corroding Western liberal democracy. A much debated study of the Trilateral Commission called it "lack of purpose" already more than ten years ago. It is as if the ideological fountains of liberal democracy, shackled by its intrinsic tolerance even to those bent to destroy it and inhibited from positive action by granting every individual and nation the inalienable right to choose their own way of life and fulfillment, had struck bottom; as if its spiritual roots had gone stale and are no longer capable of spearheading the urgent innovative political thrusts, so needed to help adapt our archaic world order to the exigiencies of change and a different global arrangement. Addressing this problem, *Michael Crozier* wrote that the only thing the cultural vanguard of Western Europe—torn between the spectre of a Soviet takeover and a nuclear blowup—has brought forth is an "apocalyptic nihilism." He concluded—valid also for the rest of Western democracies—that "no model of civilization emerges from the present-day drifting culture, no call for reform and pioneering."[3] With regard to the U.S., one is tempted to add that, except for its preoccupation with balance sheets and productivity rates, it lacks most a genuine vision for the future. Its spiritual radars, as I have pointed out earlier, are more tuned to the preservation of traditional structures and values, than to the winds of change rocking its political foundations. This perhaps explains the long record of political blunders of the U.S. and other leading Western democracies in their dealings with the Soviet Union over the past forty years, their helpless dabbling with the Third World problem, and their pathetic weakness in handling, what *Yehezkel Dror* calls, "crazy states" and "crazy movements," among which international terrorism occupies a prominent place.

No wonder that the reputation of liberal and representative democracy in Third World countries, where stable democratic institutions and dedication to the common good are even sparser rarities than in the North, is far from unblemished and that its pretension to provide necessarily the best course for underdeveloped nations, is seriously questioned.

Capitalism, as Marx admitted more than a hundred years ago, permits prodigious development of productive forces. To unscrupulous entrepreneurial aggressiveness, colonialism, fierce competition, and innovation the world owes much of its technical progress and material well-being. The motor of this process, self-interest and profit, has boosted man's knowledge and wealth, given him unprecedented power over the forces of nature, and built empires. Yet in its present form of state-capitalism, run by monstrous transnational corporations, controlling the world economy and generating unchecked technological revolutions at a wizardly pace, boundless consumerism and Third World stagnation, capitalism and the creed upon which it is founded face increased criticism and rejection.

One basic target of attack is that modern capitalism fosters "uncontrolled pursuit of economic objectives . . . limitless economic growth at the expense of social progress." The inevitable consequences are the deterioration of our ecosystem, diminished quality of life, the proliferation of deadly weapon systems, the boom and bust cycles, shaking the world economy, and bankruptcy facing many Third World countries. Analyzing the roots of this development, *Daniel Bell* holds that "American Capitalism . . . has lost its traditional legitimacy, because it has substituted Protestant sanctification of work . . . with hedonism which promises material ease and luxury."[4] The hypnotic preoccupation with growth rates, accumulation of wealth, and status-seeking—the torch and banner of a whole epoch—seems to have made capitalist societies increasingly insensitive to social effects and needs, not only in their own countries, but especially in the Third World. Because its basic rationale is the naked profit motive, capitalism, as *Robert L. Heilbroner* has stressed, is uncapable of creating an "ethic of solidarity and cooperation," a cause so desperately needed for overcoming the chasms dividing the world.

But critics accuse capitalism of an even more outrageous crime—it does not and probably cannot produce truly egalitarian systems. Even in the U.S. and the Western Euopean countries, the disparities between a dazzlingly rich minority and a very poor stratum at the bottom of the population pyramid are painfully conspicuous. In the affluent OECD countries there were over twenty-five million unemployed in 1982. After 200 years of existence, capitalism is still unable to guarantee man the right to work, though it is consecrated in Article 23 of the Universal Declaration of Human Rights.

But capitalism has also flunked on the global scale. Except in a small number of nations, it has failed in the majority of Third World countries to spark anything remotely comparable to the dynamic growth processes and material and social progress attained in the U.S. and the other Western industrialized nations. Even in countries endowed with rich natural resources such as Mexico, Chile and Venezuela, it has only succeeded over the past 150 years to create economic torsos and to perpetuate patterns of abysmal income disparities between small wealthy elites and the masses living in abject poverty. Capitalism just doesn't seem to possess the right remedies for reversing the widening North-South gap. Its market mechanisms for free trade and financial and technological transfers, tend to make rich countries richer, while the poor stay poor. Unfortunately, unabashed looting by multinationals in cahoots with grabby local elites is still the general pattern. For all its fantastic economic and technological performance in the industrialized North, capitalism tends to create and perpetuate unequal economic relationships between social classes and between nations. It is therefore ill-suited and inadequately equipped for dealing with some of the most crucial issues confronting mankind at this historical crossroad.

Soviet Communism, as a political system, has a few advantages over Western democracy. Its top decision-making processes and policies are not subject to questionable obstructionism from opposition platforms or to an ill-informed or witch-hunting public. The Politburo decides. The custom of communist leaders to perpetuate themselves in power gives the Soviet regime, since Stalin, considerable stability and continuity. It has no problems with destabilizing phenomena as political dissent, strikes, subversion, and social violance.

But the price the Russian people have to pay for these dubious advantages is high: Total subjugation of the individual to the Moloch of State and Party, absence of freedom of expression, control by sinister secret service apparatusses, and the reality of the Gulag for political dissenters, described so impressively by *Solzhenitsyn.*

The era of Stalinist mass liquidations, purges, and mock trials are gone. But during the long reign of *Brezhnev, Khrushchev's* short-lived thaw was stopped, *Dubcek's* Prague Spring was snuffed out with Warsaw Pact tanks and the Solidarity movement in Poland suppressed on order from Moscow. Communism with a human face has remained a dream, in the Soviet Union, Cuba, and Vietnam. Whether *Gorbachev's* battle cries of *glasnost* (openness) and *perestroika* (transformation) will go that far, is very much open to doubt. It is debatable, admits the Yugoslav philosopher *Svetozar Stojanovich,* whether the achievements of the Communist regimes have been worth the tremendous human cost, "the pyramids of sacrifice," in *Peter Berger's* words.[5] The lesson is clear, as long as its brutal disregard for the individual persists and it keeps supporting revolutionary violence and world revolution, Soviet-style Communism is a fear-inspiring alternative and its causes remain as suspect and unacceptable as the methods it uses for attaining them.

As a gospel, Communist ideology still rings revolutionary bells, except in the Communist nations themselves. In the Soviet Union, Marxism-Leninism is a spent force. Ideology, says the Polish philosopher *Leszek Kolakowski,* "only has the function to legitimize the monopoly of power (by the Communist Party)."[6] It is, as a high State Department official has been quoted, "the Soviet version of compulsory school prayer." That Marxist orthodoxy rings like dead leaves was felt by the fathers of the bold Czechoslovakian attempt to revitalize the mummified Communist structures and dogmas with a democratic blood transfusion; by the French, Italian, and Spanish Euro-Communists, who renegaded on such doctrinarian taboos as "the dictatorship of the proletariat" and "the necessary violent seizure of power"; and by *China's Teng Hsiao-ping,* who realizes that many of its tenets are out of touch with today's realities. *Marx's* pivotal thesis on the progressive impoverishment of the working class in capitalist countries and the fading away of the state lie in shreds and—worse yet—*Enrico Berlinguer* dared to lift the accusing finger and declare that "the Soviet Union could no longer be considered a model for the construction of socialism." Though ossified historical and dialectical materialism may still be taught at the Moscow State University, there is little doubt that Soviet brand of socialism and utopian Communism are export labels with decreasing resale value. Communist political theory and practice are bound to be even less valid and appealing in the world of tomorrow without cheap labor, run by computers and with higher standards of living. But whether the Soviet Union, pledged to continue fighting the class enemy, supporting the national liberation struggle and aiming at a Communist world dictatorship, can afford to throw all this obsolete ideological ballast overboard without endangering the survival of its political elite and the system itself, is the big question.

There is but one area left where the political and ideological seeds of Soviet Communism still fall on fertile ground—the underdeveloped world. It thrives on the poor accomplishments of mock democracy in many of these countries and on the infatuation of their leftist avant-garde with the socialist principle of state ownership of the means of production and the perspective to shoot themselves into positions of unrestricted and perpetual power. But these possibilities, as I have shown in an earlier chapter, are waning also. *Regis Debrais* already called attention to the virtual demise of Communist parties in many African and South Asian countries and to the fact that "in Latin America (outside Cuba) there are less militant Communists in 1985 than in 1945."[7]

Soviet Centralized Planned Economy. According to Marxist-Leninist doctrine, socialist economies, run in strict accordance with short and long-term plans and unhampered by profit motives and costly competition, should perform miraculously well. But reality is different. In spite of maintaining relatively steady growth rates, a high pace of industrialization and a slow but continuous rise of standards of living, the Soviet economy—seventy years after the October Revolution—is still a poor second runner to the Western market economies in terms of efficiency, productivity, and standards of material well-being.

Hamstrung by unimaginative bureaucracies, lack of incentives and stimulating competition and harassed by unrealistic plan targets and chicaneries of Party apparatchiks, it works in general with the sluggishness of a cranky and unoiled motor. While in the U.S., 4 percent of the working force feeds the nation, producing besides huge surpluses for export, in the Soviet Union, 30 percent can't do the job. Yearly imports of U.S. or Argentinian wheat have to guarantee Iwan Iwanovich his daily bread. In industry, Soviet technology and productivity rates avowedly lag far behind Western levels. The Soviet R & D performance, though backed up by twice the number of scientists and engineers employed in the U.S., is still no match to the innovative creativity ebullient in capitalist America, Japan, and Western Europe. *Nikita Khrushchev's* flamboyant predictions on the Twenty-first Party Congress in 1959 that the new Seven Year Plan (1959–65) would be "the decisive step . . . to catch up and surpass the most developed capitalist countries in per capita production," and that, "already in 1970, and maybe earlier, the Soviet Union would occupy the first place in the world as far as total production and per capita output are concerned," have been quietly forgotten. Truth rings a different story. "In real growth," writes *Prof. Folke Dovring* from the University of Illinois, "the Soviet Union is now falling behind many capitalist countries and may also be falling behind the world as a whole."[8] Accepting *Lenin's* dictum, that superiority in productivity will decide the final outcome of the contest between socialism and capitalism, Soviet socialism has already flunked that test. *Chairman Gorbachev's* scathing attacks against the drowsiness, complacency, and inefficiency of the system's economic elite; indiscipline of workers, and irresponsible and wishy-washy attitudes of unions; and his desperate and almost revolutionary reform course of *perestroika,* are the best and the most irrefutable barometers of the Soviet economy's stagnation and profound ills. The conclusion is evident. The editor of *der Spiegel, Rudolf Augstein,* stated it bluntly: "A country can't be governed with Leninist methods. *Lenin's, Stalin's* and *Gorbachev's* state is not competitive."[9]

Socialist and communist systems are superior to the capitalist counterpart in one field only—they tend to be more egalitarian, they do away with extreme poverty, procure to give everybody work and provide people with a certain standard of living. However, the standard of Ivan Ivanovich is still leagues below that of John Smith in the U.S. or any of his Western European or Japanese pendants. As to the alleged superior Soviet social benefits, it suffices to recall the impressive catalogue of *Andrei D. Sakharov,* which shows that they still fall far short from the benefits granted the average citizens in the West. *Djilas's* basic thesis in his classic, *The New Class,* has still not been refuted. The privileges of the *nomenclatura,* Mercedes limousines, lovely "dachas," and access to capitalist luxury articles in shops, barred to the common Russian, contrast shockingly with the vexing shortages of basic supplies and housing ordinary people have to put up with.

The egalitarian gospel is a seducing theme also in Third World countries, where in quarters of the poor and power-hungry leftist intellectuals it probably exerts more appeal than the West's esoteric, often rather hollow catch-phrase of "individual freedom." In the face of capitalism's poor performance in most of these countries, the same holds true with regard to the principle of "central planning." But as the record of almost all socialist experiments has demonstrated to satiety, this principle promises more than it can hold.

Soviet-style socialism or communism has also failed in creating a new culture and particularly in creating the "new man," the dream of *Trotsky, Mao,* and *Ché Guevara.* In the Soviet Union, and in most other socialist countries, Western consumption and culture patterns such as Coca-Cola, jeans, music styles, and art trends enjoy a fascinating bonanza. The low working morale, corruption, alcoholism, bigotry, and spreading of bourgeois values, lambasted by *Gorbachev,* shed a quite revealing light on the myth of the "socialist man," that cannot be argued away with rockets orbiting Earth, Russian dominance in chess and world records in weightlifting.

If we now try to draw a final conclusion from this brief critical analysis of Western liberal democracy and capitalism on one hand, and some of the basic political and economic aspects of Soviet communism on the other, a rather disconcerting picture emerges. Both systems have strong and weak points. Each is superior to the other in some aspects und inferior in others. Neither has the answers for our global crisis. Neither may lay the monopoly claim for offering the best option for man's future. If asked for a choice—and with an eye on these reservations—I prefer the fresh air of individual liberty and initiative, alive in the Western world to the Gulag atmosphere in Communist countries. I also think that the Promethean creativity and vitality of the capitalist system is something worthwhile retaining. On the other hand, Socialism does better dealing with poverty and creating a more egalitarian society and it may be better suited for promoting development in Third World countries than the unscrupulous market system.

It is no coincidence that the idea, if only both systems might fuse into one, conserving the valid components of each and jettisoning the others, has been broached by scientists time and again. If the West could curb its excessive materialism and bring the discriminating market forces under control, and if communism would shed its inhuman totalitarianism and abandon its plans for world conquest, an eventual homogenization of both systems, the end of confrontation, and beginning of a world living in peace and harmony, might be possible.

Proponents of the "convergence theory" hold that this process has started already. Nobel laureate *Jan Timbergen* argued that due to rapid industrialization and evolution of a specialized state burocracy in the Soviet Union, growth of well-being, and increased East-West cooperation, social structures of both systems were becoming more similar. The sociologist *Pitrim Sorokin* went even further, sustaining that from 1930 to 1960, not only the economic system but also social institutions, ways of living, and culture and value systems of East and West had grown more alike and that in due course of this converging trend a new world society would emerge—neither capitalist nor Communist.

However, in the wake of the collapse of detente and escalation of the conflict between the two superpowers in the late seventies and early eighties, convergence lost plausibility. *Zbigniew Brzezinski,* President Carter's top foreign policy adviser, pointed out that two systems with such diametrically opposed fundamental political objectives and methods of achieving them, could never really gravitate toward identity. Though the spectre of a looming nuclear showdown seems to have receded since the amiable tête-á-têtes between *President Reagan* and *Chairman Gorbachev* in Geneva and Reykjavik, no marriage of convenience has been ushered in by the two leaders. All fancy illusions are misplaced. The real purpose of *Gorbachev's* liberalization drive is not to introduce liberal democracy but to keep socialism from falling further behind capitalism, and to make it worldwide into an even more formidable force. He has not relinquished the Soviet Union's long range goal of world domination. Nor has *President Reagan* slackened his determination to continue standing up against the "empire of evil," and pressing the crusade for a revitalized capitalist America. What we witness today is not the peaceful merge of capitalism and socialism into *Sorokin's* "optimal society," but a highly competitive tug-of-war not only for self-preservation of each system, but for imposing it on the competitor and the rest of the world, with plenty of built-in time bombs for peaceful or not peaceful defeat of the other side, but defeat in any case.

But let us take a closer look at the future potential of both systems, which is really all that counts. In a previous chapter I have argued that the biggest challenge of our age is not how to avoid nuclear holocaust or to diminish the development gap, however imperative the search of solutions for these dramatic issues may be, but how to overcome the division of mankind into two basically irreconcilable camps and how to advance toward a global one-world-order, which would permit mankind to do away permanently with the all-encompassing crisis cluster for good. What perspectives do both systems offer with respect to these crucial goals?

Paradoxically enough, both gravitate toward culmination of this irreversible historical process. In the West, political unification of the world is perceived as a moral mandate and a task already ripening in the womb of today's scientific and technological revolution. As the *Second Report to the Club of Rome* and the report of the *Brandt Commission*—among many others—have stipulated, the world's global problems need global approaches and solutions. True world peace, as *Carl F. von Weizsäcker*—among many other prominent thinkers—has pointed out, is only possible in a world state. Soviet communist doctrine says very much the same. It holds that only after the final defeat of the class enemy and capitalist imperialism, and a successful communist world revolution, men will be able to establish a superior rational and peaceful world order. Fundamental tendencies in both systems are in tune with the most prodigious demand of civilization—its unification in a single global order. But there are profound differences with regard to the desired final results and the methods for attaining them.

A totalitarian or federal world state could, as it has been argued, very well be the upshot of an all-out nuclear clash between the two superpowers and their allies. It is probably a fair guess that rebuilding civilization would require joint efforts on an unprecedented global scale. Only world government would be up to such a task. It is hardly a farfetched conclusion that should the Soviet Union triumph, it would waste no time and make *Lenin's* dream of a communist world dictatorship a quick reality. Since communism, as *Jean-Francois Revel* puts it, "considers itself permanently at war with the rest of the world," the men in the Kremlin are unlikely to miss this chance, after having brought the archenemy, the United States, and hated capitalism, down to their knees. If, conversely, the U.S. should prevail, it too would be in the position to bring the remaining shambles into the fold of a supreme global authority, possibly a federal world state along lines compatible with Western political philosophy.

Which of the two alternatives is more desirable is a matter of political gusto. As of now, neither has all the credentials. However, it is safe to assume that after the nuclear dust has settled, neither Western democracy nor communism would be the same. Facing a worldwide, post mortem Nuclear Winter scenario, America's political system would have to acquire strong egalitarian and solidary features and introduce global planning on an unprecedented scale. Vice versa, a Soviet communist world dictatorship, with no more class enemies and imperialists around, would necessarily soften up and eventually undergo a certain democratic metamorphosis.

It is a totally different question, considering the poor performance of the Soviet economy, whether communism would measure up to the global task. The exiled Czech philosopher *Julius Tomin* once predicted that "the occupation of Western Europe would destroy the Soviets from within." If he is correct, then *Gerd von Hassler's* quip, "Imagine, communism would have to fulfill its hundred-year-old promise of creating paradise on Earth," acquires more than rhetorical meaning.[10] Nevertheless, a communist world government—though its socialist order would lack capitalist drive and efficiency—is a possibility whether we like it or not.

Oddly enough, the same cannot be taken for granted in case of a U.S. victory. *André Malraux* once remarked, "Americans lack an imperial style," and *George F. Kennan* admitted that "America's possibilities . . . as an active partner of world affairs, do not lie on the global plane."[11] Nothing could be more true. The U.S., and the West, in general, lack ostensibly a clear-cut vision of the world that might emerge in the aftermath of a nuclear conflagration. They have not the slightest ideological or organizational blueprint for organizing it. "Communism," says *Jean-Francois Revel,* "is a better machine for world conquest than democracy." No matter how ruthless, it may also be a better machine for organizing the world thereafter. The United States, as I have pointed out earlier, let its enviable lead after World War II slip away with the lofty carelessness of a bohemian count of the eighteenth century, unaware of the social turmoil knocking at his door. "Despite its formidable economy," writes *Richard J. Barnet,* "its network of military installations, stretching around the globe, and its self-proclaimed mission to guide internal political developments in the Third World with military interventions if necessary, the United States lacked the inner confidence either to press for victory over the Soviets or to develop ground rules for long-term coexistence."[12] But the increase of Soviet might and prestige was due to more than Washington's pathetic complacency. The Kremlin pursues its professed goal of world domination with more determination and purposefulness. In comparison, the U.S. has nothing more exciting to offer than *President Reagan's* unctuous declaration that "the United States has the gift of democracy and that it wants to share it with the rest of the world." It all boils down to one basic weakness: The U.S. has been and still is terribly short of a global sweeping political design. This, more than anything else, gives Soviet communism, which suffers no weakness of this kind, a significant edge. The capacity of the West to provide enlightened leadership for the necessary profound reorganization of the world's global order in the hypothetical case of a Soviet defeat must therefore be viewed with considerable skepticism.

But the solution to stalemate and global confrontation need not necessarily come via all-out war. There is still a good chance that sanity and the will to compromise will win the day. The possibility that the U.S. and the Soviet Union may agree on withdrawing their short- and medium-range nuclear missile arrays on the European theater may be a first valuable step in this direction. It may be followed by other even more important steps furthering disarmament and leading eventually to a substantial reduction of the strategic nuclear potential of both superpowers. But a healthy dose of doubtfulness is not misplaced. The *Khrushchev–Kennedy* peace talks in Vienna in 1960 had a brutal awakening in the Cuban missile crisis only two years later. *President Carter's* peace overtures to *Brezhnev* found their quick Waterloo in the Soviet invasion of Afghanistan, and *General Jaruzelski's* coup in Poland. A similar fate may yet await the deescalation attempted by *President Reagan* and *Chairman Gorbachev.* Neither of the two is Superman. In *Horst E. Richter's* excellent fiction parody *Everybody Spoke about Peace,* visiting extraterrestrials, trying to reconstruct why the human species had decided to end it all by committing a collective suicide, discover that "its statesmen had been mostly exhausted and sickly figureheads, shaken by a multitude of compul-

sions to take decisions, a task that none of them had been equal to."[13]
The parallel with our reality couldn't be more evident. *President Reagan,* shaken up by Iranscam, may just be grabbing disarmament and rapprochement as the last available straw to remake his image of a great statesman with a major breakthrough on the foreign front. *Chairman Gorbachev,* fully aware of the backwardishness corroding the Soviet economy and society and facing dogged resistance from the apparatchiks to his attacks on ancient dogmas, may just be baiting for the necessary time of respite and aiming to build an even more colossal machinery for world conquest. Neither of the two objectives would seem to come to grips with the paramount issue facing our divided and crisis-shaken civilization and bring us within grasp of the needed global solutions.

Trust is missing between the two opposing camps, and to create it requires more than lusty handshakes, off-the-cuff exchanges of platitudes, and the submission of proposals that are knowingly unacceptable to the other side. To rekindle a genuine atmosphere of trust, it will require the firm commitment of both sides to reconsider their basic positions, the will to deescalate the decisive conflicting issues and, above all, deeds to prove it. It is here where the real difficulties begin.

In order to show its good will, the U.S. could ask the Soviet Union to abandon its goal of "world domination" and to stop supporting puppet regimes and revolutionary movements in the Third World, representing a constant menace to world peace and balance of power. For the same token, the Soviet Union may demand from the U.S. to withdraw its nuclear umbrella from the NATO theater in Europe, to forget about Star Wars, and to cancel out its far-flung strategems for encircling Russia. Evidently such demands would be as difficult to fulfill at this stage as the U.S. proposal that the Soviet Union should permit free elections in its orbit, or the Soviet request that the U.S. should return Alaska, Oregon, and part of California to Russia.

Though not reluctant to agree to a new disarmament deal, Washington has shown no signal that the West's basic ideological, political, and strategic principles and interests may be negotiable. Neither is the Pentagon prone to junk its grand schemes for absolute military superiority and to substitute them with amiable teas with Soviet marshals and showy hit parades. On the other hand, the Soviet Union is far from being China. In China, *Teng Hsiao-ping's* heartening pragmatism has launched a billion Chinese onto a road of development that has buried most of the sterile ideological squabbles with the West, and that is rooted in the refreshing insight that "the classic Marxists are incapable of solving many of China's foremost problems" and that "a bit of capitalism can not harm China if it helps overcome the country's backwardishness faster."[14] There is neither talk nor are there attempts on the part of Peking to take over the world. Soviet immobilism, ingrained resistance to political and economic reform, and a false sense of pride exemplify exactly the opposite—a morbid reverence of the obsolete structures and ideological myths of Soviet socialism and communism. In spite of *Chairman Gorbachev's* alluring overtures, there are no indications so far that Soviet communism is about to scrap its face and throw its long-term goals overboard. Boulder-sized obstacles as these help to explain why rapprochement, true detente, and deescalation in stages is such a difficult, if not hopeless, task. When *Alexander Solzhenitsyn* asked the Soviet leaders in 1973, in a letter that was widely publicized in the West, "to discard communist ideology" and "to renounce unattainable missions of world domination," this was asking water to start running upstream.[15] *Andrei Sakharov's* proposals for "convergence of the two systems," contained in his famous manifesto of 1968, in which he propounded for the first phase of rapprochement such Soviet concessions as "a multiparty system," "the right of Soviet republics to secede," the "freedom to strike," and "a profound liberalization of the economic system," were even more unrealistic. Opponents of Reaganism may just as well have demanded from the U.S. Congress to curtail individual freedoms and anathematize private property. The absurdity of such proposals hardly needs further comments.

But couldn't it be, optimists may interject, that *Gorbachev's* bold reform course, the rehabilitation of *Andrei Sakharov* and *Alexander Solzhenitsyn,* and the Kremlin's apparent intention to withdraw the Red Army from Afghanistan are the kind of deeds signaling a genuine change of heart on the part of the Soviet Union? Wouldn't an agreement between the U.S. and the Soviets to do away with their short and medium-range nuclear missiles in Europe be a firm indication that both powers have begun to search seriously for some kind of arrangement and means of coexistence with each other? If this is the case, if the summit meetings in Geneva and Reykjavik marked the initiation of a first stage, with more substantial concessions and agreements to follow, then the bipartite dialogue would indeed merit worldwide acclaim and possibly mark the beginning of a new "live and let live" global relationship between the two powers and opposing systems. But so far, Geneva and Reykjavik have only lit hope on a tiny candle that may prove just another disappointment.

It might have been different. If both sides had really been willing to come to terms, they might have brushed the hardline approach aside and started delineating, at least in rough contours, a broad outline of a common strategy for peace and lasting coexistence. *President Reagan* and *Chairman Gorbachev* could have startled the world with an inspired statement, declaring their decision to end the ghastly East-West confrontation and their willingness to underwrite the fresh start in their relations with a package of far-reaching mutual concessions. To prove the sincerity of their intentions, they may have given the solemn pledge that in subsequent stages all major obstacles, blocking until now true progress in this direction, would be placed on the negotiating table in a spirit of mutual restraint and compromise. The agenda would include defrosting stymied disarmament talks, German reunification together with a gradual dismantlement of NATO and the Warsaw Pact, a political solution to the grueling Near East conflicts, the Soviet pull-out from Afghanistan, the stop to Soviet subversion in Central and South America and other parts of the world, and a moratorium on SDI. To give this historic agreement the necessary punch, they might have gone a bit further, creating a mixed top-level study group with the task of mapping out the guiding lines for the gradual implementation of this grand peace strategy, with phases for reducing tensions in all principal conflict and danger areas, and establishing the ground rules for mutual action with regard to such sensitive issues as international terrorism, crazy states, genocidal practices, destabilizing economic bankruptcy, and famines. Instead of continuing in their conniving schemes for technological, military, and strategic superiority, both sides may have gone on considering further political and ideological concessions, fostering bonds of economic and scientific collaboration and cultural exchange, and widening ever more friendly and peaceful competition and the scope of aid to the Third World. In the longer perspective, even more ambitious goals, concerted efforts to avert the plight of the LDCs, joint enterprises for space colonization and exploration, and a gradual overhauling of our anarchic international order, including the feeble UN system, may have been visualized.

Regretfully, no such path-breaking proposals, much less agreements, have been forthcoming from the Geneva and Reykjavik summit meetings. The naive expectations that the *Reagan–Gorbachev* get-togethers could have turned the helm around and steered the ship of civilization into the harbor of peaceful coexistence of the two opposing systems, have so far remained little more than idle intellectual daydreaming. None of these proposals were, of course, within the reach of practical feasibility. Neither of the two leaders were prepared or authorized to commit themselves to such revolutionary goals. The systemic roadblocks, built into the mummified East-West conflict, only permit peripheral advances, step by step, at a snail's pace, if at all.

They vouch that the devilish rationale, motoring this conflict, keeps nourishing distrust and hateful rivalry, and perpetuating the global deadlock ad infinitum. "In its essence," admitted the former East German professor *Wolfgang Seiffert,* "the basic conflict between the two existing political systems—of the Soviet type and that of the Western world—is unsurmountable."[16]

In his vision of the year 2081, the physicist *Gerald O'Neill* ventures the optimistic guess that life a hundred years hence will be rosy and full of handsome rewards. But today's basic tendencies point rather in the other direction. In *This Endangered Planet,* author *Richard Falk* predicted that in the eighties, "the policies of desperation," and in the nineties, "the policies of catastrophe" would be calling the tune.[17] Maybe in the wake of mutual concessions on disarmament, nuclear catastrophe in the proximate future can be averted. But we would be well-advised to remain on guard against a lighthearted optimism. Reykjavik has not and could not usher in the great turning point—if anything, it made the ironbound character of the global confrontation more transparent. One final delusion may just as well be cast away. A liberalized and economically more efficient Soviet communism, led by the young and dynamic *Gorbachev,* will be an even tougher competitor for the easily tricked U.S. and a more formidable pretender of the scepter for world domination.

Other pioneering ideas should therefore be explored, capable of helping to create currents of thought transcending the petty chauvinisms of both systems, and furthering the development of a true world consciousness. This consciousness would have to be deeply rooted in the insight, that lasting solutions to man's global problems will escape him, unless he unites in one big brotherhood of nations, in which, as *René Dubos* put it, "our ability to create the equivalent of the typical tribal unity that existed at the beginning of the human adventure,"—though on a much higher level—will be tested.[18] The idea of extraterrestrial intelligence, of its search and contact with it, encapsulates such a cosmopolitan mobilizing message. It rises above the political and ideological dividing lines of our distraught world order and may provide man's imagination and hope with new stimuli that things must not remain as they are, that the establishment of a peaceful and juster global order, rich in material and spiritual rewards, is by no means a utopia. Knowledge that other intelligent beings have managed to overcome their intestinal wrangles and unite in a superior global organization would impart the electrifying spark that is still missing today. After that, I think, nothing on Earth will be able to remain the same.

PART TWO
ETI AND OUR STALEMATE

Chapter 11

The Basic Proposition

In the preceding chapters I have touched upon some of the principal aspects of our contemporaneous worldwide embroglio. The portrayal does not pretend to be complete. Many important facets have been sketched only with scanty paint. Others, such as the energy crisis, have been barely dealt with. Though a general trend of recovery marks most Western industrialized nations, the world economic crisis is far from over, nourishing the suspicion that its basic ailment may be of a much deeper nature than anticipated. The conspicuous incapacity of the developed countries to heave the derailed train of the world economy back on its rails, highlighted by structural upheavals and high unemployment rates in their own backyards, adamant Third World stagnation and the precarious state of the international financial system, is a particularly dangerous symptom of this decade. The still skyrocketing and chaotic foreign debt situation of most developing countries exerts effects of a delayed time bomb that may rip the shaky international finances wide open, and blow the basic contradictions between the "have" and "have not" nations right into our faces. The population crisis, too—though demographic growth rates have begun to slow down in some developing countries—is still a galloping nightmare, swinging a cruel cudgel at many of the poorest regions of our globe. It has already wiped out whatever little progress in per capita rates these countries had been able to achieve in past decades, and now threatens to accelerate prospects of even greater human calamities and drama in the near future. International terrorism is another sickening scourge, still met by most afflicted nations with an astonishing degree of tolerance and impotency. Finally, there is also the heavy toll of brazen violations of human rights, not only in totalitarian dictatorships, but even in many countries masquerading as democratic, especially in the form of brutal racial discrimination, ethnic liquidation, and religious fanaticism.

The most alarming feature of this situation is man's pitiful incapacity to measure up to its inherent challenges. Dismal crisis-management, if not crisis-escalation, has become the professional ersatz of statesmanship-like crisis-solution. As I was able to show repeatedly, the rationale underlying the East-West deadlock functions as a ruthless clutch, holding humanity's fate on the course of self-perpetuating stalemate and crisis-magnification. All other facets of our crisis-syndrome spin almost mechanically according to the rules of its intrinsic logic.

This grisly state of affairs may drag on well into the twenty-first century. Nuclear stalemate, people agree, is better than nuclear holocaust. The underlying hope is that given enough time, the East-West confrontation is bound to deescalate into a new period of detente and U.S.—Soviet cooperation. "We are . . . more likely to ascend triumphantly," predicted *Herman Kahn* in one of his last books, "than to be condemned to eternal damnation."[1] Addressing the French Parliament in October 1985, *Chairman Gorbachev* referred himself to Hamlet's meaningful saying "To be or not to be," and brandishing a conciliatory olive branch, he expressed—for a Communist leader—the unusual conviction that the survival of civilization "can be assured only if we learn to live together, to get along on the small planet by mastering the difficult art of showing consideration for each other's interests."[2] Since then, *President Reagan* responded to this placating approach with some conciliatory rhetorics and proposals of his own, strengthening speculations that *Voltaire's* dream of the ultimate "triumph of reason" may, after all, gain the upper hand. But even if the present disarmament talks should yield tangible results, probabilities are great that we will have to go on living with the volcanic nature of world politics, local wars and violence, turmoiling social conflicts, Star Wars saber rattlings, cutthroat technological competition, and parochial fanaticisms for a long time. In his evolutionary path from primitive animal to the dimension of a mature human being, man is still related to his fellow man as an enemy and competitor and not as a brother and friend.

As long as these circumstances prevail, tendencies favoring the development of a rational global conscience, of a *Weltanschauung*, drawing its strength from the conviction that man can only find peace and harmony in an organizational entity, transcending the nation-state, will have a hard time to assert themselves. Acceptance of nonpartisan universal values and nonsectarian outlooks and political options will keep meeting stiff opposition not only from extremist nationalism, still rampant in many Third World countries, but also from doctrinarian dogmatism and escapist mysticism, still revered by conservative forces in East and West. In the European community, efforts are under way to create and promote the image of the new "European Man." But this regional endeavor, however valid, is not enough. Our time calls desperately for the creation of "universal man," man of planet Earth, with deeply anchored allegiances not only to one color, tongue, or flag, but to the cause of all mankind.

Search of extraterrestrial intelligence and contact with it, I believe, can help humankind advance toward this supreme goal. Efforts of science to this end cannot be a substitute for the urgent necessity of people everywhere to keep pressing for change and the reshaping of the existing world order. We cannot afford to fold our hands and wait for superior beings from other worlds to come to our rescue and teach us how to survive. *Carl Sagan* rejected this haywire notion a long time ago, pointing out that "the expectation that we are going to be saved from ourselves by some miraculous interstellar intervention, works against the necessity for us to solve our own problems."[3]

Contact with a superior civilization will change completely the course of mankind's history. It will give us access to an amazing wealth of scientific, political, and cultural information that would inevitably entail revolutionary impacts on man's thought, his technology, economy, and life. In due time, it may also influence our ways of relating to each other, individually and collectively, and spur man's determination to move on to a higher form of global organization.

But the most important aspect of the proposition on ETI, as I see it, is not what contact—once the cosmic dialogue gets rolling—may report to man in terms of new scientific knowledge and technological insights, however breathtaking in scope and repercussions this cosmic know-how transfer may be in the long run. Of even greater significance is its mobilizing potential for us today. "The world," wrote *Walter Sullivan* more than twenty years ago, "desperately needs a global adventure to rekindle the flame that burned so intently during the Renaissance, when new worlds were being discovered on our planet and in the realms of science."[4] The admonition still rings, with more vibrant chords than ever. Search for other intelligent beings in the universe, I think, provides this thrilling adventure.

SETI-efforts encapsulate a message pregnant with meaning for our crisis-stricken globe. The scientific supposition that technologically and organizationally advanced civilizations are likely to coexist with us in galactic space, burns the mandate into mankind's conscience to stop his mad race at the brink of apocalypse and start unifying hearts and souls of people everywhere on behalf of the causes of the human species as a whole. "The greatest significance of the Pioneer 10 plaque," insists *Carl Sagan,* "is not as a message to out there; it is a message to back here."[5] Striking a similar note, *Isaac Asimov* also underscores that SETI "can't help but emphasize the pettiness of our own quarrels and shame us into more serious attempts at cooperation."[6]

Search for ETI implies therefore much more than the desire of astronomers and astrophysicists to listen in on the cosmic murmur of faraway worlds. It is a proposition with implicit far-reaching perspectives for the future of *Homo sapiens*. That, so far, this has not been adequately appreciated and that the subject matter has often been presented in a frivolous and distorted way forms part of the moldy sarcophaguses, in which much of popular mentality and bias is cast. In trifling science fiction, the technically highly advanced, but politically and morally backward and evil extraterrestrial is still a widely exploited and believed stereotype. It is conveniently forgotten that the two paths of technical and social evolution are parallel, intimately linked processes, joined at the very seams like the two molecular strands in the double helix structure, and that this is probably a common trait of all extraterrestrial intelligent life. Civilizations which are highly advanced scientifically and technologically are necessarily equally advanced socially and politically.

Search for and contact with extraterrestrial intelligence is therefore a human enterprise, going way beyond the realm of natural sciences. But until now this enterprise has remained largely the private domain of a distinguished but small elite of natural scientists and engineers. The 112th Symposium of the International Astronomical Union (IAU) on *"The Search for Extraterrestrial Life: Recent Developments,"* held in Boston in June 1984, underlined this paradoxical situation. Contributions centered on the technical aspects of ongoing search projects, on new, sophisticated techniques for receiving cosmic signals, the feasibility of interstellar travel, planetary evolution in the galaxy, and the probabilities of complex life forms evolving on stellar systems different from ours. But nothing or hardly anything was said with regard to a number of highly provocative issues and questions of a nontechnical nature, as the probable longevity of advanced civilizations, whether they are likely to be benign or evil in our moral sense, and if contact with such a civilization would set off a demoralizing culture shock here on Earth, as it has been postulated by a few writers.

Another set of incisive questions has been posed by *Carl Sagan*. "We know who speaks for the nations," he stated, "but who speaks for the human species? Who speaks for Earth?" In case of contact, who will answer and what should we reply? In case of an approaching extraterrestrial spaceship or probe asking for permission to enter our space or to land, who is to decide and how? What are the effects, the advantages and disadvantages that we may expect from contact with advanced extraterrestrial beings? Last but not least, how could it be avoided that information obtained from them will not be usurped by one of the world's powers for its own selfish interests, and assured that it will be used for the benefit of all mankind?

Search for extraterrestrial intelligence and contact with it is not an undertaking only involving huge antennas, spectrum analyzers, and intricate biological experiments. It is a world-unifying cause, and as such, it is the task of both, natural and social sciences, to show that all people on this globe have something valuable and vital to gain from it.

Put in a slightly different way, search for ETI and contact must stop being portrayed as thrilling intellectual exercises, something permitting our imagination to delve into mysterious worlds of beauty and perfection and revel in the delicacies of a "Fourth Encounter" honeymoon with immortal creatures resembling butterflies, as gratifying as such flights of imagination may be. The endeavors of SETI and CETI have to break out from the ivory tower, where select minorities cultivate it, and become a vast human enterprise.

Search for extraterrestrial intelligence is where it is today thanks to the breathtaking advances of natural and engineering sciences. The crests of the truly revolutionary waves are ridden by the pioneering feats in molecular physics and microelectronics, in biochemistry and genetics, in radioastronomy and rocketry. It is here where the trailblazing ideas and breakthroughs are born, where constantly greater futures are generated by conquests of new scientific and technological frontiers. It is now up to the humanities to discover the global dream of ETI and to guide it along with the hard sciences to contact and beyond.

As *Giuseppe Cocconi,* from CERN in Geneva, admitted in his introductory address to the IAU Symposium in Boston, "culturally SETI had, and still has, some peculiar difficulties in being accepted by the public opinion."[7] In many intellectual circles ETI is still a suspect matter. The reason for this skepticism is only partly rooted in *Jack Catron's* caustic conclusion that "there cannot be an intelligent civilization on another planet for the same reason that there isn't one here."[8] According to *Occam's* Razor, probably a much simpler explanation accounts for this skeptical attitude. Most people still ignore or possess only the scantiest information on what the scientific proposition of ETI is all about. The following chapters are an attempt to address this shortcoming.

Chapter 12

Search for ETI: An Impressive Record

Our civilization is approaching rapidly the advanced state of science and technology that will allow us to answer one of the most fundamental questions in Nature—How common is Life, and especially advanced life, in the universe?
—Michael D. Papagiannis

In 1959, *Giuseppe Cocconi* and *Philip Morrison,* two American physicists, proposed in *Nature* that searches of signals of extraterrestrial intelligence should be conducted on the 21 centimeter wavelength of hydrogen. Since hydrogen is the most abundant element in the universe, they assumed that advanced extraterrestrials would use its wavelength for messages beamed at other civilizations. A year later, in 1960, the astronomer *Frank Drake,* from the National Radio Astronomy Observatory (NRAO) in Green Bank, launched Project Ozma One. Over a period of 200 hours, he attempted to detect, with a one-channel receiver, interstellar radio signals from two neighboring stars, Tau Ceti and Epsylon Eridani, about twelve and eleven light-years away from our sun. Shortly after, in November 1961, a conference was held at Green Bank under the auspices of the National Academy of Sciences, where *Frank Drake* proposed his much discussed equation $N = R_* f_p n_e f_l f_i f_c L$* for calculating the number of technologically developed civilizations in our Milky Way galaxy.

These three scientific landmarks highlight the takeoff point of the debate on the probable existence of extraterrestrial intelligent life, that has gained scope and interest ever since. Long before, other in-

*N—the number of technologically advanced civilizations;
R_*—rate of star-formation during lifetime of galaxy;
f_p—fraction of stars having planetary systems;
n_e—number of planets ecologically suitable for life;
f_l—fraction of planets on which life actually occurs;
f_i—fraction of planets on which intelligence arises;
f_c—fraction of planets on which intelligent beings develop a communicative phase;
L—mean lifetime of a technical civilization.

quisitive minds, among them *Titus Lucrecius Carus*, the Chinese philosopher *Teng Mu, Giordano Bruno*, the French writer *Camille Flammerion*, and the Croation-born U.S. inventor *Nicola Tesla*, speculated that other worlds and living beings must exist in the universe. But only the *Cocconi-Morrison* contribution, *Ozma One*, and the *Green Bank Conference* transformed the speculations on ETI in a practical working hypothesis and nailed SETI down on the dissecting table for scientific analysis and experimentation.

About the same time, similar initiatives regarding ETI were spurred on the part of Soviet scientists. In 1963, *Josip Shklovskii*, a leading astrophysicist from the Sternberg Astronomical Institute in Moscow, espoused in the little booklet *Universe, Life, Reason* some challenging ideas on ETI, stirring great interest in the Soviet Union. At the All-Union Conference on "Extraterrestrial Civilizations and Interstellar Communication," in Byurakan, Armenia, in 1964, the attending astronomers and physicists agreed that scientific advances had boosted the probability that intelligent life exists elsewhere in our galaxy and that communication projects over stellar distances were now technically feasible. *Shklovskii* suggested that most extraterrestrial civilizations are probably more advanced than ours, and *N.S. Kardashev*, from the Institute for Cosmic Research in Moscow, proposed to categorize all civilizations in the following three classes:

Class I—with a technology level and energy output close to the present-day level on Earth;
Class II—civilizations which have harnessed the energy output of their star;
Class III—civilizations harnessing the energy output of their entire galaxy.[1]

Shortly thereafter, *Kardashev* stunned astronomy circles and world opinion with the exciting news that he and his colleagues had filtered out artificial signals from an astronomical object, called CTA 102. But the announcement turned out to be a false alarm.

Interest in SETI now began to pick up also in the international scientific community. A CETI study group of the International Academy of Astronautics (IAA) met in Madrid in 1966, proposing studies on interstellar communication problems and the potential impact of contact on mankind. Only a year later, the IAU set up a committee for organizing an international symposium on ETI. It was held in September 1971 at the Byurakan Astrophysical Observatory in Yerevan, Armenia, sponsored by the Academies of Sciences of the U.S. and the Soviet Union. The meeting, assisted by fifty-four top scientists and engineers, mostly Americans and Russians, permitted a thorough cross-examination of astronomical, physical, chemical and biological issues relative to ETI and came up with a series of significant resolutions, reproduced here in Appendix A. Conviction among the participants that eventual contact would bring mankind great benefits was high. However, their proposal to establish an international working group on ETI failed to get official blessing in Washington and Moscow.

An important event was the launching of the *Pioneer 10* spacecraft from Cape Kennedy in March 1972 with the mission of exploring the planet Jupiter. The small sonde carried a gold-anodized aluminum plaque with a message, devised by *Carl Sagan* and *Frank Drake* and dedicated to intelligent extraterrestrials, containing some basic scientific information, a clue regarding our whereabouts, and a representation of a man and a woman, greeting their likes in space. The sending of this plaque, followed by an identical copy on *Pioneer 11*, launched in 1973, focused worldwide attention as man's first intentional message to other intelligent beings in the cosmos. Because of the sondes' slow speed of about thirty kilometers per second, it will take them some 80,000 years to come to the vicinity of Alpha Centauri—the closest star to our sun—a mere 4.3 light-years away. The likelihood that they will ever be intercepted by a space-faring advanced civilization, and that the plaques will be deciphered, is therefore probably very small. But the sending of the plaques ratified man's desire to break out from isolation on Earth and to make his existence known in this part of the galaxy.

Already in 1971, scientists of the NASA Ames Research Center presented *"Project Cyclops,"* proposing to construct at a cost of $5 to 8 billion an array of over a thousand radio telescopes, arranged in a circle of some 20 km^2, with the aim of carrying out searches of artificial extraterrestrial signals from a couple of million stars up to a thousand light-years away. But the project never received the green light. The Soviets, in turn, finished building in 1976 in the Caucasus the giant radio-telescope "RATAN 600" with 14,000 m^2 of aluminium reflectors, set in a circle of 600 meters and interconnected with the main radio-astronomical centers of the Soviet Union—to be used, inter alia, for SETI projects.

The next important step forward was the transmission of a message by the radio-telescope in Arecivo, Puerto Rico, in November 1974 to the globular cluster Messier 13 at a distance of 25,000 light-years. This message in binary code, using the format of black-and-white television, was the first interstellar message of Earth, traveling at the speed of light. Its content included basic physical, chemical, and biological data, a drawing of our planetary system, and the sketch of a telescope, indicating the stage of our technology. Some scientists, among them *Sir Martin Ryle,* a Nobel laureate, objected to the message, on the ground "that it was very hazardous to reveal our existence in the galaxy." The majority, however, celebrated the feat of *Frank Drake* and *Carl Sagan* as a triumph of our communication capabilities with cosmic space.[2]

From 1972 to 1976, the astronomers *Benjamin Zuckerman* of the University of Maryland and *Patrick Palmer* of the University of Chicago, carried on the ambitious search project *Ozma Two.* Some seven hundred stars up to a distance of sixty-five light-years were analyzed. Similar projects were conducted in the Soviet Union at the Sternberg Institute of Astronomy and the Gorky Institute of Radiophysics. Neither of these projects was successful. Notwithstanding, the Soviet Academy of Science approved in March 1974 an ambitious research program on CETI. Search projects were now proliferating. In the U.S., the most outstanding were conducted by *Robert S. Dixon* at Ohio State University, by *Verschuur* at the NRAC, who zeroed in on ten nearby stars, and by *Sagan* and *Drake,* who directed their receiver at whole galaxies. Similar projects were initiated in Canada and Australia—all with negative results.

In the meantime, studies had also been pushed in the field of space travel. The British Interplanetary Society presented in 1978 the results of *Project Daedalus,* which showed that with a reasonable extrapolation of today's technology, an unmanned flyby of *Barnard's Star,* at a distance of about six light-years, will be feasible in the later part of the twenty-first century. Though interstellar travel is still fraught with prodigious technical and economic problems, this landmark study bears witness to man's faith that one day in the future he will take his spaceships up to the stars.

NASA now also began to take an active role in SETI. Its report, "The Search for Extraterrestrial Intelligence," of 1977, prepared by scientists of the Ames Research Center in California, revealed that NASA had conducted several workshops on the subject, that it considered SETI research "both timely and feasible," and basically "an international endeavor in which the U.S. can take a lead."[3] As a follow-up, the Jet Propulsion Laboratory in Pasadena, received the go-ahead from NASA for a five-year program to survey 80 percent of the sky, though funding later reduced the program to 250 stars.

Of great significance were the launchings of *Voyager I* and *II* in August and September 1977, scheduled to explore the outer planets of our solar system. Attached to each *Voyager* craft, which since then returned a gold mine of scientific information and thousands of spectacular pictures, is a practically indestructible gold-coated copper phonograph record, containing fascinating messages to intelligent extraterrestrials; including 118 photographs of our planet and our civilization, a selection of some of the world's greatest music, an audio with meaningful sounds of Earth and well-wishes in fifty-five languages together with special greetings by *President Jimmy Carter* and the Secretary General of the UN, *Kurt Waldheim.** This remarkable feat illustrated the uninterrupted scientific and technological progress in this field, which has made possible sending exploratory robot stations and recorded messages, endurable for billions of years, deep into planetary and stellar space.

Interest of national and international institutions in ETI had now been fully awakened. In the U.S., the Subcommittee on Space Science and Applications prepared for the House of Representatives a report in 1977 on the "Possibility of Intelligent Life Elsewhere in the Universe."[4] The ITU commissioned its Study Group 2, "Space Research and Radioastronomy," to include SETI in its coverage, while the UN Committee on the Peaceful Uses of Outer Space placed the subject on the agenda of its Scientific and Technical Subcommittee.[5] A research group of the European Council also jumped on the bandwagon late in 1979 with a contribution on the "Evolution of Intelligence and SETI."

Increased outlays for the development of improved observation facilities were now forthcoming. They facilitated, among other things, that the dream of astronomers to mount a telescope in space, free of interferences, will come true when the European Spacelab, which will be mounted inside the shuttle orbiter, will be equipped with the large optical telescope (LOT), under development by NASA. This new telescope is scheduled to serve also such SETI projects as the detection of planets in other stellar systems, of extraterrestrial artifacts in interplanetary space, and of huge colonization and space exploration enterprises. The Soviet Union too, plans to use a space-born radio telescope, to be delivered by the *Kosmoljot,* the Soviet pendant of the shuttle orbiter, to the yet-to-be orbited space station, capable of supporting twenty inhabitants round the clock, at least part time for searches of ETI.

Curiously enough, as the seventies drew to a close, a skeptical attitude regarding ETI suddenly began to branch out in some U.S. scientific and governmental quarters. In 1978, the Appropriation Committee of both Houses cut NASA's request for a two million dollar SETI program, and though the scientists testifying at the first U.S. Congres-

---

*Appendix B.

sional Hearing on SETI favored strongly continued federal support of NASA efforts in this field, the Golden Fleece Award was given by *Sen. William Proxmire* to the modest NASA-SETI project for 1979. The same senator killed the project once more in 1981, irked by reports that NASA had continued to support low-scale SETI research surreptitiously.

This temporary reversal for NASA's SETI efforts, partly due to the downward swing of the U.S. economy in the late seventies and early eighties, had its roots also in the fact that after the first period of exaggerated expectations regarding the detectability of ETI, a more sober trend had begun to set in in scientific circles. Disappointed by the fact that not one of the search projects conducted so far had been able to provide positive evidence, some scientists expressed doubt, wondering whether the early optimistic appraisals about life and intelligence in the galaxy didn't need to be revised.

Though in the course of this trend, *Josip Shklovskii*, once an ardent believer in ETI, backpedaled, adopting a pessimistic view on the existence of other intelligent life, the official Soviet SETI projects were advancing full swing. At the SETI symposium, "Communication with Extraterrestrial Intelligence," sponsored by the Soviet Academy of Science in Tallinin in December 1981, *V.S. Trotskii* disclosed that "a new search system was to be built in the Soviet Union especially for SETI, utilizing one hundred antennas of 1 meter diameter each, allowing continuous observation of all the celestial sphere above the horizon."[6] Other Soviet participants advanced details on the construction of the $75 million seventy-meter parabolic telescope near Samarkand, equipped with an instrumental capacity exceeding any other telescope in the world, on the projected installation and operation of telescopes in space and on launching antennas into high orbits, to be used partly for SETI purposes.

Recalling the *Apollo-Soyuz* rendezvous in 1975, informal talks were held at this symposium on possible future joint U.S. and Soviet SETI programs. But due to the strained relations between the two countries and the doldrums suffered by NASA's SETI project, nothing came of the proposition.

However, the setback experienced by the official SETI program in the U.S. proved to be shortlived. Early in 1983, NASA's SETI program was reinstated by Congress, thanks to a firm recommendation from the Field Committee on Astronomy, but probably also influenced by an international petition, signed by seventy leading scientists from many countries, among them nine Nobel laureates.* This reversal en-

*The text of the Declaration is reproduced in Appendix C.

abled the Ames Research Center and the Jet Propulsion Laboratory in Pasadena to concentrate on developing highly sophisticated techniques for computerized search programs, as the development of spectrum analyzers having up to 10 million individual frequency channels, compared with the 74,000 channel prototype available until then. The listening ability of SETI receivers are to be increased this way by a factor of several million. Some eight hundred stars, similar to our sun, are to be examined. In addition, other modes of observation, using the thirty-four-meter diameter antennas of NASA's Deep Space Network of satellite tracking stations of three sites around the world, are supposed to permit scanning the entire sky.[7] Also in 1983, project *"Sentinel"* was launched at the Harvard University/Smithsonian Astrophysical Observatory by *Prof. Paul Horowitz.* Similar to the SETI program, conducted at the Ohio State University, it initiated a round-the-clock all-sky search for artificial signals.

In this long string of important SETI endeavors, the creation of *Commission 51—Search of Extraterrestrial Life* at the eighteenth General Assembly of the IAU in August 1982 was a particularly significant event. It decided to engage in a broad sphere of scientific activities, many related with SETI,* and it is now held that because of its impressive membership of 240 prominent astronomers and other scientists from 28 countries, Commission 51 marks the take-off point of SETI as a genuine international enterprise.[8]

It was also encouraging news that in its 1982 report to the National Academy of Sciences on *"Astronomy and Astrophysics for the 1980s,"* the Astronomy Survey Committee recommended "an astronomical Search for Extraterrestrial Intelligence," and that it commended the allocation of twenty million dollars for work in this area.

All this shows that the transient obstructions which blocked concerted official SETI efforts in the U.S. for three years, have been surmounted and that fresh air is oxygenating new, more systematic, and multifaceted approaches in this fascinating field. The breathtaking advances achieved over the past three decades were probably best highlighted at the IAU Symposium 112 *"Search for Extraterrestrial Life*—Recent Developments," which took place in Boston in June 1984. Attended by 150 renowned scientists from eighteen countries, it permitted not only a complete updating of the respective state-of-the-art, but also marked a milestone, legitimizing SETI as an important and serious scientific undertaking. In his concluding remarks, *Prof. Michael D. Papagiannis,* president of the IAU Commission 51, stated,

*The objectives of Commission 51 are in Appendix D.

"All in all, it is fair to say that the search for extraterrestrial life and intelligence has finally come of age."[9]

This fact that SETI has reached internationally a scientific standing of unquestionable distinction is born out by the impressive record that emerged from this symposium. "In the past twenty-five years that have passed since the first SETI project," *Professor Papagiannis* disclosed, "there have been close to fifty searches . . . radio observatories in Australia, Canada, France, Germany, Holland, the USSR, and the U.S. have participated in this effort," and so far, "close to 120,000 hours of observations have been logged." Though none of this projects presented evidence of the existence of ETI, it should be kept in mind that all had been using still comparatively primitive technics. Most participants at the symposium agreed that only with the operationalization of the new projects, using fabulously sophisticated devices, SETI will reach the intensity and scope that is required for a solid assessment of abundance or scarcity of intelligent life in our galaxy.

One of the most exciting projects is the development of NASA's multi-channel spectrum analyzer (MCSA), which will be able to resolve a wide band of frequencies simultaneously into as many as 8.2 million individual channels. It will be used in NASA's SETI program, which, as *Bernard M. Oliver,* head of the SETI project at the Ames Research Center, explained, "combines two strategies: a targeted search of late F,G, and early K stars in the solar neighborhood and an all-sky survey, comprising about 1000 targets . . ."[10] The search is scheduled to be fully operational by 1990.

Just as promising looks *Project META*, the successor of Project Sentinel at Harvard University and sponsored by the Planetary Society, whose 8.4 million channel receiver began operating in September 1985. Helped along by a contribution of $100,000 by *Steven Spielberg,* producer-director of *E.T.* and *Close Encounters of the Third Kind,* the Megachannel Extraterrestrial Assay improves Sentinel's capabilities nearly a hundredfold, increasing greatly the possibilities that the programmed search in the coming decade will hit paydirt.

Other promising SETI projects are *Robert Dixon's* endeavors at Ohio State, which detected two areas near the galactic center, emitting still unexplained strong narrowband pulses, the improved SERENDIP II project at the Space Science Laboratory in Berkeley and a search conducted by the Department of Astronomy of the University of Washington, which concentrates on eavesdropping on stars for leaking radio signals. Japan has also initiated SETI activities, by focusing its forty-five meter radio telescope of Nebeyama Radio Observatory on promising stars in front of dark clouds.

In the Soviet Union, search projects now seem to be favoring surveys in the infrared to microwave regions of the spectrum, on the assumption that super-civilizations would surround themselves with Dyson-spherelike structures, emitting in this range. The seventy-meter telescope under construction near Samarkand is likely to be used for this purpose.

One of the most striking aspects is the revolutionary pace in technical advances of SETI, which—as scientists agree—will make a thousand Ozma One searches per second possible. But progress in SETI-related fields, as in technology for detecting other planetary systems and the discovery of micro-organisms of extraterrestrial origin, will also open up promising perspectives.

Two significant trends have asserted themselves thanks to this prodigious string of advances. One is the unflinching conviction of highly prestigious scientific circles that, regardless of negative search results so far, the rapidly expanding knowledge in microbiology, planetology, and cosmology make the existence of extraterrestrial intelligence an even more solid and thrilling hypothesis. The other is wide agreement, expressed, for instance, in the international petition, signed by distinguished scientists in 1982, that "the only significant test of the existence of extraterrestrial intelligence is an experimental one."[11]

It could be held that official support of SETI in the U.S. and the Soviet Union is partly due to the rivalry between the two superpowers and their incurable fear that the other side may achieve a decisive technological breakthrough either through contact with ETI or—more likely in the short run—via a militarily applicable spin-off of SETI R & D. But the main reason is that because of its scientific and technical merits, SETI now stands on its own feet as a respected discipline and because, as the Astronomy Survey Committee stressed in its 1982 report to the NAS, "It is hard to imagine a more exciting astronomical discovery or one that would have greater impact on human perceptions than the detection of extraterrestrial intelligence."[12]

The case for SETI looks even better, if the extraordinary dynamics of today's scientific and technological revolution are duly considered. Decisive breakthroughs are expected to occur soon in planetology, bioastronomy, and radioastronomy, which are likely to extend man's reach into the cosmos and buttress SETI's main proposition. At the same time, a new possibly more realistic but still visionary consciousness of the importance of space and space conquest, paralleling that of the Soviet Union, is alive in the United States. In spite of the tragedy of the *Challenger,* such projects as the launching of the Infra-Red Astronomy Satellite (IRAS) and programs as Galileo and the Space Tele-scopes are bound to keep the U.S. in the space race. *President Reagan's go-ahead to NASA in his 1984 State of the Union message, "to develop a permanently manned space station, and to do it within a decade,"* highlights the determined thrust of the U.S. in the realm of space exploration and colonization, though the timetable of this project, in which Western European nations are participating, will suffer minor changes. The search for extraterrestrial intelligence can only profit from this spirit of leaving the cozy shelter of planet Earth and turning space conquest into a lasting adventure of mankind.

Chapter 13

Probabilities that Extraterrestrial Life and Intelligence Exist: The Stand of Science

> *No great discovery has ever been made in science except by one who lifted his nose above the grindstone of detail and ventured on a more comprehensive vision.*
>
> —Albert Einstein

In the preceding chapter I have described some of the major high points of the search for extraterrestrial intelligence during the past three decades and I have tried to show in a sketchy way what is going on in this new scientific discipline, in which some of the most fascinating and meaningful research anywhere is taking place. Now I will turn to the very gist of the proposition on ETI, to the question: What are the basic scientific assumptions underpinning the claim that extraterrestrial intelligence very likely exists in our galaxy and in other galaxies? What hypothesis or observational evidence do astronomers such as *Frank Drake* and *Michael D. Papagiannis,* physicists like *Philip Morrison,* and molecular biologists like *Cyril Ponnamperuma* base themselves on, by sustaining that this likelihood is great?

The first very strong argument in favor of the existence of ETI is a purely statistical one. *Isaac Asimov* sums it up this way: ". . . in the observable universe, there are as many as 1,000,000,000,000,000,000,000 (a billion trillion) stars"; and he concludes, "This one consideration alone makes it almost certain extraterrestrial intelligence exists."[1] Greek and Chinese philosophers couched their faith in the existence of other worlds and life in very similar terms. It simply eclipses our imagination that out of such an inconceivably large number of stars, stretching out in space billions of light-years, intelligent life should have arisen solely on one of the planets of our sun. But even abstracting other galaxies and considering only ours, the Milky Way system, we are still stuck with a mind-boggling number of stars, estimated at 100 to 300 billion. The proposition that intelligence should have evolved just on one of so many stellar systems, seems illogical and difficult to accept.

150

This argument, however, is more than statistics. It is related to what astronomers call the "cosmological principle," which states that except for some local peculiarities, no part of our universe is unique or privileged, and that therefore our solar system is not a special, but rather a quite typical case. The inescapable conclusion to be drawn from this principle is that the evolution of intelligent life on Earth cannot be considered an exceptional chance happening, but that it must have also developed elsewhere in solar systems similar to ours.

The second line of argumentation is closely linked to the spectacular advances in basic sciences. So it is now generally accepted that the same laws of physics which we observe on Earth also rule stellar processes and motions in other parts of the cosmos. "The laws of nature," says *Carl Sagan*, "are the same everywhere. Not only do the same chemical elements exist everywhere in the universe . . . distant galaxies revolving about one another follow the same laws of gravitational physics as govern the motions of an apple falling to Earth, or *Voyager* on its way to the stars."[2] With the constituent parts of the cosmos and processes as heat, radiation, and volcanic activity being very much the same everywhere, the likelihood that only on planet Earth living matter and intelligence should have evolved can only be infinitely small.

Of paramount importance in this relation are the scientific insights and findings regarding the multiplicity of planetary systems, the likely origin of life, and the existence of extraterrestrial living matter.

Until about thirty years ago, astronomers believed that planetary systems like that of our sun are extremely rare phenomena in the galaxy. Planet Earth with its exuberant life forms was held a unique case. Due to recent advances, especially in spectroscopic observation, this view has been radically overhauled. "Modern astrophysical and astronomical theory," summarizes the NASA scientist *John Billingham,* "predicts that planets are the rule rather than the exception and likely to number in the hundreds of billions in our galaxy alone."[3] Planets are today considered natural byproducts in the process of star formation, as hot dust clouds condense into the nucleus of a new star. The computer simulation, done by the American planetologist *Stephan Dole* in 1970, parting from a cloud of dust and gas and providing for random motions and effects, resulted in the formation of planetary systems amazingly similar to our own. Since 1977, observational evidence supports the hypothesis that planets are the rule. Astronomers of the University of Arizona are convinced that in the stellar object MWC 349 of the *Constellation of the Swan,* planets are in the process of formation, as a new star is born in the center of a disclike cloud of hot gas and dust. The detected wobbles in the trajectory of *Barnard's star* and of a few other stars are also indications that they are circumnavigated by planets. But the most exciting news came from IRAS, launched in orbit by NASA in 1983. Its three-meter telescope "was able to detect envelopes of particles around a considerable number of young stars, possibly indicating planetary formation for preplanetary discs."[4] Vega was the first such star. According to a NASA spokesman, "This discovery provides the first evidence that solid objects of substantial size exist around a star apart from our sun."

Even more fascinating discoveries are expected with the new observational instruments now under development. The Hubble space telescope, due to be launched in the U.S. in 1987, is supposed to have the capacity to detect planets directly with highly sophisticated devices, while the Space Infrared Telescope Facility, an orbiting fifteen-to-twenty-meter telescope, will be able to detect objects the size of Jupiter in nearby stars. A similar capability will be achieved soon by the Allegheny Observatory and even more promising perspectives may be opened up by the permanent space stations, already in the planning stage in the U.S. and Soviet Union.

It is therefore taken for granted that the new observation facilities will soon give us absolute certainty that other planetary systems, similar to our own, exist in our vicinity. If planets, as the majority opinion holds, turn out to be the most abundant objects in the Milky Way galaxy, it will give a powerful shot in the arm to the proposition that life and intelligence cannot possibly be unique phenomena of Earth.

Another strand of relatively recent breakthroughs has convinced scientists that the emergence of life on Earth cannot be regarded as a fortuitious and therefore necessarily rare occurence. Quite the contrary, most molecular chemists and microbiologists are convinced today that—given enough time—the evolution of inanimate and intelligent life forms is almost inevitable. Stressing the uniform physical and chemical properties of the universe—water is probably one of its most common compounds—they feel quite sure that processes as those on Earth, which led to a surprising multiplicity of life and to intelligence, must have taken place on many other planetary systems. In the superb book, *Life in the Universe,* ten eminent natural scientists, writing on SETI, summed up current theory of chemical evolution and origin of life this way: "Given a suitable and sufficiently enduring environment, life is a normal, natural consequence of the long-term application of the basic physical and chemical processes of the universe."[5] Referring himself to the astonishing fact that all living matter, plants, animals, and humans use the same organic chemistry, and even the same amino acids and genetic code, *Cyril Ponnamperuma,* former director of the Laboratory of Chemical Evolution of the University of Maryland, corroborated that "The alphabet of life is . . . extremely simple; the wide variety of life today may be traced to a mere handful of chemicals," drawing the conclusion that "if conditions similar to those on the primitive Earth prevail elsewhere in the universe, the same kind of events might happen, resulting perhaps in a kind of life very similar to ours." The evolution of life on Earth was therefore not a cosmic chance, but merely the consequence of a series of propitious conditions and interactions.

Exactly how organic life evolved from inorganic matter is still not perfectly understood. But at least parts of the mystery have been lifted already. After Earth had cooled down sufficiently, some 3.6 to 3.9 billion years ago, first living microorganisms developed relatively soon from inorganic substances in what scientists call the "primordial broth," under the influence of sunlight, heat, ultraviolet radiation, electrical discharges and volcanic activities. The respective pioneering work was done by the British biologist *John B.S. Haldane* and the Soviet biologist *Aleksandr Oparin,* who suggested already in the 1920s that the original atmosphere on Earth, when first life developed, had probably been quite different from the atmosphere of nitrogen, oxygen, and water vapors that evolved later, due to the development of green plants and of photosynthesis.*

How the first steps in the long chain of evolutionary events that led to life forms may have occurred has been demonstrated by *Stanley Lloyd Miller* in 1953. He exposed a mixture of methane, ammonia, and water to electrical discharges for a week, and when he analyzed the resulting broth, he found, much to his surprise, that substances basic to our life—complex molecules of glycine and alamine, two amino acids occuring in proteins, had formed. Subsequent experiments with simulated primordial atmosphere mixtures invariably resulted in the formation of the same and other basic building blocks of life, a strong indication of how these substances, fundamental for all our life forms, may have developed. More recent research concerning the development of more complicated macromolecules has shown that it cannot be chance that is working in their formation and that the *Darwinian*

*Oparin and Haldane developed their work independently from each other. Oparin suggested a primordial atmosphere of methane, ammonia, and water, while Haldane's was composed of nitrogen, carbon dioxide, and water vapors. It is now believed that life may have started in Oparin's atmosphere, which later—through ultraviolet radiation—gave way to Haldane's composition.

principle of self-reproduction via constant mutation and natural selection is involved in it. How self-reproduction started is still one of nature's most guarded secrets. But the wall guarding it is getting thinner fast. The fact that all living matter has the same genetic code is a strong indication that it must have developed from a small common original ancestor. Other recent research bolsters the claim that externally induced mass extinction, possibly triggered by comet showers, which may eventually have effected the dinosaurs, should not be seen as disasters, but "may actually be essential," as *J. John Sepkoski, Jr.,* holds, "to ensure the continuation of evolutionary experiment and the further development of complex life."[6] The crater landscapes on our moon, on Mars, and on other moons observed so far in our solar systems, support the view that similarily disastrous, but in the end fruitful, developmental processes must also have occurred on many other worlds, where life forms have evolved. Most scientists in the field today therefore believe that the evolution of always more complex self-reproductory and self-regulatory living forms, leading continuously—gradually and by leaps—to more complicated life forms, and eventually to intelligent life, is a natural and necessary consequence, once the chain of the process has been initiated.

A third promising line of research deals with existence of living matter outside Earth. Though Mars, earmarked by bio-astronomers as the most prospective candidate, has proved sterile so far, science has struck gold in related fields and has been able to compile an impressive record of evidence, backing up the thesis that life is not a monopoly of Earth.

Recent research in astrochemistry shows that interstellar space is full of molecules and chemical compounds well known to us. "Since 1968," disclosed *Cyril Ponnamperuma,* "many of the molecules that are important to chemical evolution, such as ammonia, water, hydrogen cyanide, and carbon monoxide have been found."[7] Information handed to the IAU symposium in Boston in 1984 showed that the list of interstellar molecules in dark nebulae, where processes of star and planet formation have probably taken place, has already increased to sixty. The evolution of interstellar particles, made up mainly of elements like oxygen, carbon, and nitrogen in combination with hydrogen into complex organic molecules is consequently a scientifically established fact. Recent evidence also seems to indicate that organic molecules may be present in the Jovian atmosphere. On Titan, the largest moon of Saturn, *Voyagers I* and *II* discovered not only nitrogen and argon and lots of hydrocarbons, but also other biochemically interesting compounds. At the mentioned symposium in Boston, scientists of the Laboratory for Planetary Studies of Cornell University sustained that on Titan, "a complex organic chemistry should be occurring."

Evidence for the existence of organic matter on other celestial bodies are also the so-called *carbonaceus chondrites,* rare and easily crumbling meteorites. The analysis of the *Murchison meteorite,* that fell in Australia in 1969, yielded eighteen different amino acids, the fundamental structural units of proteins. "Six of them," *Isaac Asimov* notes, "were varieties that occur frequently in the protein of living tissue."[8] Fatty acids were also found. These results, confirmed by recent findings of similar meteorites in the Antaractic by Japanese scientists, have given new weight to the *panspermia hypothesis,* proposed first by Swedish physicist *Svante A. Arrhenius* late in the last century. He proposed that life on Earth may have been imported from space via living spores. For decades, scientists had scoffed at this idea, arguing that such spores, even if encapsulated in protective crusts, could not survive the hostile environment of protracted interstellar journeys. But recent research seems to provide evidence that in dark clouds, where the penetration of ultraviolet light is considerably reduced, life expectancy of living spores may extend to several tens of millions of years, enough of a life span to carry life from one stellar system to another. Life on Earth, some scientists hold, may therefore go back to the parental molecular cloud.

A highly exotic version of *planned panspermia* has been put forth by the British Nobel laureate *Francis Crick* and *Leslie E. Orgel* from the Salk Institute for Biological Studies in La Jolla, California. They proposed that life on Earth owes its existence to the bombardment with living organisms which a higher intelligent species, possibly faced by extinction, directed to our planet for the reason of preserving life in the universe. This unproven hypothesis, smacking of "biological von Dänikenism," however, has been found full of unanswered loopholes. Not only should scientifically and technologically highly advanced extraterrestrials be able to avoid extinction; even more questionable seems the proposition, that—after having reached such a high stage of development—they should start panspermia on the lowest level of microorganisms.

Independent of the suspect *Crick-Orgel* version, the relevant point is that the panspermia hypothesis probably will marshal more scientific merit in the future. This development would corroborate further the supposition that life as we know it on Earth cannot be unique in the universe.

A firm conclusion can therefore be inferred from all that has been said so far. Impressive evidence from a wide field of scientific disciplines lends support to the proposition that life forms, similar to ours on Earth, must have sprung up on many of the multitudes of planetary systems now believed to populate the Milky Way galaxy. But another important question now needs clarification. What are the chances that they may have developed into intelligent life forms? Not too long ago, the view prevailed that intelligence is the private endowment of man alone. Today most molecular biologists and chemists as well as anthropologists agree that intelligence is a natural evolutionary stage of life, as natural as the evolution of life itself.

The quality of man's intelligence may, of course, be questioned. Many distinguished scientists have posed lately the perturbing question whether self-destruction and degeneration may not spell the inevitable doom for the human race and fasten a mocking epitaph onto the grave of *Homo sapiens*. But there can be no doubt that intelligence, among other attributes, gives a species an added survival value in the relentless struggle of the fittest. Cognitive capacities, foresight, and the ability to learn from experience gave man, since the early days of the *Proconsul,* some twenty million years ago, a decisive advantage over other species and vis-á-vis, a periodically hostile natural environment. It was the development of his brain, above all of the neocortex, the seat of intelligence, associated with deliberation, congnitive anticipation, logical thinking and planning that allowed the human race to establish itself as king over the myriads of animal species.

It should be a cause of meditation, though, that it took intelligence such an awfully long time to evolve. Whereas the first primitive life forms developed rather quickly about 3.5 billion years ago, only about a billion years after planet Earth was formed, our closest predecessor, *Australopithecus,* who possessed a low-degree intelligence, emerged only about 3.5 million years ago, and *Cro-Magnon man,* who precedes modern man directly, just entered the evolutionary stage about forty thousand years ago. True, from then on, up to today's computerized and atomic space age, the pace of man's development has been spectacular. But the knotty fact remains, that for intelligence to enter the scene—lets assume with the *Proconsul*—it took about 99.5 percent of our Earth's total age. *Carl Sagan's* cosmic calender, covering the period from the "Big Bang," some fifteen billion years ago, till today, illustrates the point very vividly. Making this enormous stretch of time fit into one earthly year, *Sagan* shows that while the formation of Earth occurred on September 14, and the origins of life may be dated back to September 25, the *Proconsul* and the first signs of intelligence only developed at 1.30 P.M. on December 31.[9] The cave paintings in Europe only occured at 11:59 P.M., and the birth of Christ at 11:59:56 P.M. of the last day of the year. *Michael D. Papagiannis* states the obvious by pointing out that "getting life started appears to be a much easier task than its subsequent evolution to an advanced technological civilization, which in our case took about ten times longer."[10] But though intelligence developed late in the puzzling zigzag course of biological evolution, highlighted by mass extinctions and dead-end alleys of some species on one hand and the emergence of always new and more resistant species on the other, it developed nevertheless into a quite natural form, and with the evident objective to increase our species' survival chance. In the course of the struggle of the fittest in changing conditions of the habitat, spurred by the need to adjust to the rigors caused by natural disasters, intelligence, as *Isaac Asimov* puts it, "is more or less an inevitable development on a habitable planet, given sufficient time."[11]

A different question is whether an intelligent extraterrestrial species would also possess an advanced technology, which might make contact with it possible. Dolphins are said to have elaborate brains that may be capable of quite refined thought processes or something similar. *Freeman Dyson* thinks therefore that "it is easy to imagine a highly intelligent society with no particular interest in technology." But this is an idea hard to cozy up to. *Dyson's* proposition may hold true for intelligent extraterrestrial beings living in a medium—as the dolphins do—with abundant food supplies at their disposal over almost indefinite periods of time and, consequently, without the need to develop elaborate manipulating organs as our hands, without which the development of technology is impossible. But considering the hazards of planetary evolution, from periodic tectonic upheavals, meteorite showers, and close passes of other celestial bodies to profound atmospheric and radiation changes, this ideal case should be the exception. The case of our ancestor relative, the *Proconsul* of the late Tertiary, who was forced to perfect versatile manipulative organs in order to build for himself a successful ecological niche, in a tough and changing environment, is probably the much more typical case. Modern technology, for all its sophistication, is but the result of a process that started with a stick and a stone millions of years ago.

I think evolution of intelligence and technology is an intricately interwoven and interdependent process. Man's history is an ever-widening stream of ideas, of intellectual prowess and fertility, but it is even more so a steadily expanding wave front of always mightier and more intricate technological accomplishments. Trying to envisage the evolution of man's intellectual powers without paying homage to the uninterrupted technical revolution which gave, step by step, material expression to this ascending process, is a sterile endeavor. Inherent in intelligence is the innate urge to push back the frontiers of the unknown and to harness the natural environment in the interest of the basic needs of a given species—continued growth and security. The continuous translation of new scientific insights into always more sophisticated levels of technological performance—the case of man—is probably a sublime natural quality, written into the genetic code of most intelligent species. I can't imagine that extraterrestrial civilizations should be an exception to this rule.

It is true that so far definite proof for the existence of intelligent life on another celestial body eludes us. "The concept of subjective probability," wrote *Prof. Terrence Fine* of Cornell University more than fifteen years ago, "is at present the only basis upon which probability estimates can be made about extraterrestrial intelligent life."[12] But since then the veils of cosmic mystery, still withholding final evidence, have worn thin in crucial areas. Thanks to the greatly perfected observational instruments soon to become available, most astronomers today feel optimistic that foolproof findings on the multiplicity of planetary life in other worlds of our galaxy will soon be in our possession. Though some of the most recent findings suggest that the number of habitable planets may be less than it was thought earlier, the existence of extraterrestrial intelligence has become a formidable scientific proposition. "It appears," stated *Michael D. Papagiannis* at the IAU Conference in 1984, "that our civilization is approaching rapidly the advanced state of science and technology that will allow us to answer one of the most fundamental questions in Nature—'How common is life, and especially advanced life, in the Universe?'"[13] Subjective probability of ETI is giving way to objective probability. With the aid of the great discoveries expected before the turn of this millennium, it may well become an unassailable certainty.

Chapter 14

Number of Extraterrestrial Civilizations: Estimates and Debate

Once our society becomes convinced of the existence of intelligent life elsewhere in our galaxy, we will embark on the greatest voyage of discovery in all our history.
—Bernard M. Oliver

After presenting in a nutshell some of the main scientific arguments speaking for the existence of ETI, I may now address the question that has intrigued and baffled researchers for the past twenty-five years: How many extraterrestrial civilizations are likely to exist in our galaxy? And at what distance are they from our sun?

Distinguished scientists have devoted much effort trying to shed some light on these interrogations. The reason is obvious. If the likelihood is great that the number of such civilizations is very large, it should be possible to establish contact with the nearest one in a relatively short time. On the other hand, if estimates tend to support the opposite, then proof that we are not alone in galactic space and contact may still elude us for a much longer time.

Today, science is divided on this puzzling issue. Scientists believing that intelligent life is abundant are opposed by others holding that we may be the only intelligent race around. A third group feels that in the absence of confimative evidence, such estimates are just a waste of time. "Arguments about the (Drake) equation," quipped *Charles J. Seeger* from the San Francisco State University, "should occur only over a table with plenty of beer, but not in print." I prefer to think that estimates on the frequency of extraterrestrial civilizations are a valid endeavor, if for no other reason, because they address a fundamental question that is necessarily on everyone's mind.

The early optimistic estimates parted from *Frank Drake's* equation $N - R^* f_p n_e f_l f_i f_c L$, mentioned in chapter 11, proposed in Green Bank twenty-six years ago. Assuming the average lifetime of an extraterrestrial civilization to be ten million years, *Carl Sagan,* then an outspoken proponent of the optimistic current, estimated at the conference on CETI in Yerevan in 1971 that "there are a million technical civilizations in the galaxy."[1] This corresponds to about one in every hundred thousand stars. The distance to the nearest intelligent extraterrestrial species would be no more than a few hundred light years.

Isaac Asimov's estimate is even more buoyant. Parting from 280 billion planets in the galaxy and making subsequent allowances for such factors as a suitable sunlike star, a useful ecosphere, the factual emergence of life, et cetera, he concludes that "the number of planets in our galaxy, on which a technological civilization has developed, is 390 million," and that "the number of planets in our galaxy on which a technological civilization is now in being, is 530,000."[2] In this case, the average separation between two advanced civilizations would be about 650 light-years.

The most optimistic appraisal has been presented by the American astronomer *James A. Wertz.* From the premise that, "as the evolution of the galaxy progresses, more and more technical civilizations inhabit the galaxy," he concluded that "the number of technological civilizations . . . at the present time is 5×10^8 or 500 million."[3] Under this assumption, which would place the nearest at the doorstep of our solar system, only thirty-five light-years away, the galaxy would practically be teeming with intelligent life.

These optimistic estimates drew heavily on the conclusions of the U.S. astronomer *Stephan Dole* that there are some 650 million earthlike planets scattered throughout the galaxy. Though this number is now held to be overly optimistic, current estimates are still high. In 1980, the British scientists *Alan Bond* and *Anthony R. Martin* placed the number of planets orbiting stars of sufficient age for intelligent life to have formed, at 2.4 million, their average distance being about 180 light-years.[4]

More conservative numbers have been advocated among others by the U.S. astronomer *Sebastian von Hoerner.* Considering the potentialities for self-destruction and degeneration, he conceded in a publication of 1972, extraterrestrial civilizations a longevity of only 100,000 years, and by assuming that "about 1 percent of all stars should have habitable planets"—about 2,000 million for the whole galaxy—he concluded that "our galaxy would contain about 40,000 technical civilizations . . . half of them more advanced than we are."[5] But in a later paper, *von Hoerner* felt compelled to reduce this number to a mere 10,000, admitting that his assumptions on the lifetime of technical civilizations were still wild guesses.

Von Hoerner introduced the interesting thought that if intelligence arose on the oldest star systems hundreds of millions or billions of years ago, the first civilizations would have established contact with those evolving at a later stage, and that this process would have furthered the development of a truly galactic culture. Entertaining similar ideas, the American-Australian radio astronomer *Ronald L. Bracewell* stirred up considerable excitement with a book in 1974, in which he advocated the belief that extraterrestrial civilizations were already linked in to what he called the *Galactic Club.* He presumed that these civilizations had surely sent unmanned interstellar probes to suitable life-supporting stars as our sun, and that such a probe may well linger around in the depth of our solar system, waiting for the appropriate moment to reveal to us its existence.[6]

According to the optimistic view, often identified as the *Sagan-Drake chauvinism,* which held sway during much of the sixties and most of the seventies, our galaxy is quite populated with intelligent extraterrestrial civilizations, though disagreement prevailed as to the most probable exact number.

This view has, as of recently, been confronted by a more pessimistic trend of thought, set in in some scientific quarters of the U.S. and Soviet Union. Its main proposition, called in honor of its top proponents the *Tipler-Hart chauvinism,* and which garnered major publicity in the symposium, "Where Are They? A Symposium on the Implications of Our Failure to Observe Extraterrestrials," organized by the astronomer *Ben Zuckerman* of the University of Maryland in 1979, is that the emergence of life and of intelligent life is a much rarer phenomenon than it had been thought and that consequently *Homo sapiens* may very well be the only intelligent species in the galaxy.

One argument of this pessimistic school boils down to a significant question. If so many older and far advanced civilizations exist, how come they haven't detected us yet and made themselves known to us? Astronomers, studying the possible rate of colonization in space, have come to the conclusion that a space-faring civilization, decided to stretch out in space, would be able to colonize the Milky Way in five to ten million years. Since the galaxy is probably about twelve billion years old and intelligence is supposed to have developed on planetary systems possibly hundreds of millions, and maybe even billions of years older than our solar systems, it is a valid and puzzling question, why none of these civilizations have visited us or tried to colonize Earth. Spinning this question, called the "Fermi Paradox," after the famous Italian nuclear physicist *Enrico Fermi,* who first posed it; a bit further, *von Hoerner* postulated that "Our Earth should have been colonized long ago, and we ourselves should be the descendents of some early settlers, and not the homegrown humans that we certainly are."[7]

The physicist *Frank J. Tipler* from Tulane University also suggested that a highly advanced extraterrestrial civilization, bent on exploring the galaxy, would build "self-repairing and self-reproducing von Neumann probes," and send them to search other intelligent life forms. From the fact that no such exploratory probes have been detected by us, *Tipler* concludes, "Our technological civilization is the only civilization which has ever existed in the whole of galactic history."[8]

A different line of arguments has been put forth by the astronomer *Michael H. Hart* from Trinity University, San Antonio, Texas. On one hand, he questions the thesis that life is likely to develop on most planets with a suitable ecosphere, stating that "the probability that an arbitrarily chosen strand of nucleic acid (storing the hereditary information allowing organisms to reproduce), consisting of four types of different smaller components called nucleotides and with some 400 positions, would spontaneously arrange itself in the order needed to produce life, would be only 10^{-30} even in 10 billion years, or one chance in 1,000,000,000,000,000,000,000,000,000,000." From this, *Hart* concludes that "The probability of life arising on a given planet—no matter how favorable conditions on that planet may be—is less than one in 10^{-30}."[9] On the other hand, computer simulations run by *Hart* show that "if the Earth's orbit were only 5 percent smaller than it actually is, during the early stages of its history, there would have been a *'runaway greenhouse effect,'* and temperatures would have gone up until the oceans boiled away entirely." Conversely, "if the Earth-sun distance were as little as 1 percent larger, there would have been a *'runaway glaciation'* on Earth about two billion years ago. The Earth's oceans would have frozen over entirely, and would have remained so ever since, with a mean global temperature of less than minus 50 degrees F."[10]

Hart's simulation findings, which have been verified by other researchers, cannot be dismissed lightly. They suggest that the habitable region around a suitable G-O type star as our sun, is not a broad band, but rather a narrow fringe, bordering on calcination at the inner rim and on eternal ice on the outer edge. Other simulations have also shown that only small variations in the mass of the star and size of the planet would reduce substantially the chances of a stable ecosphere over a long period of time. On account of these new research findings, it is probably necessary to revise the original *Drake* formula, and to downgrade the early, too optimistic estimates of the number of extraterrestrial civilizations in the galaxy.

But not all of the *Tipler-Hart* criticism has gone unchallenged. Trying to explain the disconcerting *Fermi Paradox,* several theses have been advanced, among them the *Zoo Hypothesis,* holding that we have been detected a long time ago, but were considered too primitive and therefore unworthy of contact, the theory that for unknown reasons we may have been placed under some sort of *galactic quarantine* and declared *forbidden territory,* and the supposition that a *galactic code* may prohibit its members to interfere with nascent civilizations. None of these explanations strike me as very convincing. Even if extraterrestrials had visited Earth long before our time, the fascinating realm of its flora and fauna would have certainly been of great scientific interest to them, for the same reason we investigate primitive life forms on Earth and in space. Besides, they certainly would have returned periodically, to check out on the end products the vagaries of earthly evolution would lead up to. Nor does it seem plausible to suppose that contact should be forbidden, because information exchange and eventually-needed assistance between two intelligent species could only enhance the development and survival chances of both.

A much more convincing argument that may help to explain the *Fermi Paradox* has been given by *Carl Sagan.* He maintained already twenty years ago that "if contacts are made on a purely random basis, each star should be visited about once each 100,000 years"; only after contact with an intelligent species has been established, the frequency of contacts would increase, so that "each communicative technical civilization should be visited by another such civilization about once every thousand years."[11] If advanced extraterrestrials had paid Earth visits during the height of the Cretacious period, some 100 million years ago, when dinosaurs were roaming the central North American plains, they may have well concluded that evolution on planet Earth was bottled up in a hopeless cul-de-sac and crossed our solar system out as a potential site for bearing intelligent life. If, on the other hand, extraterrestrial beings—as *von Däniken* keeps arguing—had visited Earth during the ancient times when the Indian, Sumerian, Mesopotamian, Mayan, and Inca cultures flourished, they would not only have left conclusive and indestructible evidence of their visit, but they would have returned. Since no such evidence exists, it is probably safe to surmise that out of a sample of billions of possible candidates, even for a technologically highly advanced civilization, detection of a life and intelligence engendering planet is a tough proposition.

The likely reason, as it is now widely acknowledged, are the enormous interstellar distances. A round trip to Earth from a star only a hundred light-years away—a trifle in galactic terms—at the enormous speed of 0.3 c (c = speed of light), or almost 100,000 kilometers per second, would take more than 800 years, because of the needed additional time for acceleration and deceleration during the stages after takeoff and before landing. That great hazards are involved in such flights, as the tragic blowup of the *Challenger* demonstrated early in 1986, is another factor. Long flights would necessarily increase the danger possibilities. Though taking into account highly sophisticated propulsion systems for interstellar travel, longevity of the crew, and Einstein's *time dilation,* the distances for crossing galactic space may be an almost prohibitive handicap.

Carl Sagan and *William Newman* have calculated "that if a million years ago a spacefaring civilization with a low population growth rate emerged two hundred light-years away and spread outward, colonizing suitable worlds along the way, their survey starships would be entering our solar star system only about now."[12] A sphere with a radius of 1,000 light-years, which may be the distance to the nearest intelligent civilization, contains a million stars and probably several million planets; one with a radius of 5,000 light-years, several billion stars and many more planets. An ambitious colonization program could therefore be an enterprise that may very well defy the resources that even a highly advanced spacefaring civilization may be willing or able to allot to such a task.

As it is now admitted, the gigantic energy requirements, necessary for interstellar travel, may place prohibitive obstacles to daring space-exploring civilizations. *John H. Wolfe,* from the Ames Research Center of NASA, calculates that "in a round-trip powered flight to the Alpha Centauri system at approximately 0.7 c . . . a spaceship weighing about one thousand tons would require for the thirteen-year journey 33,000 tons of matter/antimatter fuel, delivering a total energy equivalent to the electrical energy consumption of the entire United States for over 300,000 years."[13] Making further allowances for the need of a shield "the equivalent of a ten meter thickness of solid tungsten," in front of the spaceship and weighing more than two million tons, *Wolfe* comes to the chilling conclusion that "a matter/antimatter fuel mass of two million tons, equivalent to eighteen million years of U.S. electrical energy consumption would be required." Boosting a huge spaceship to high relativistic speeds—a staggering and forbidding task for us at this stage—may also give an advanced extraterrestrial civilization considerable economic headaches.

Similar reasons may explain why the radio searches conducted so far have failed to detect artificial extraterrestrial signals. Maintaining sending projects over indefinite time, especially if they don't yield results, is an expensive business, provoking the axe from skeptical appropriation committees, and whose efficiency is further hampered by the fact, that the sending teams don't know which of the infinite bandwidth and frequencies another intelligent species may be able to receive, nor where to direct the beam and for how long.

Nikolai S. Kardashev attacks the *Fermi Paradox* from another angle. He is convinced that "extraterrestrial civilizations have not yet been found, because in effect they have not yet been searched for."[14] Instead of concentrating on nearby stars, he suggests, search strategies should be reoriented and concentrated on large and powerful objects and structures that may be the work of super-civilizations. Still other scientists have suggested that many extraterrestrial civilizations may refrain from colonization efforts because they are afraid the descending civilization may turn into a hostile competitor of power and living space. The theory has also been proposed that extraterrestrial civilizations may not be interested in spreading out in the galaxy and in locating other intelligent species, and that they may have decided to stick to their planet and to keep quiet. But this explanation, as I have argued earlier, is highly improbable. A scientifically and technologically advanced civilization would necessarily have an outward look and, driven by the innate knack for penetrating into the mysteries of creation, it would search for other life and intelligent life forms.

Returning to *Michael H. Hart's* extremely skeptical views, it must be pointed out that his pessimistic calculations on the probability of life arising in a suitable ecosphere are seriously questioned by top specialists in the field. At the mentioned symposium of the University of Maryland in 1979, attended by many ETI skeptics, *Cyril Ponnamperuma* upheld the thesis that "Life is considered to be the inevitable consequence of the evolutionary process in the universe." The flaw in *Hart's* argument, as *Isaac Asimov* explained in reference to similar ideas, put forth by *Lecomte du Noüy* in 1947, is to think that chance is the only arbiter in the formation of complex "protein chains, made up of one hundred amino acids, each one of which could be of twenty different varieties." "Actually," *Asimov* says, "atoms are guided in their combinations by well-known laws of physics and chemistry, so that the formation of complex compounds from simple ones are constrained by severely restrictive rules, that sharply limit the number of different ways in which they combine,"[15] The famous *Miller* experiment of 1953 did not produce a random amount of chemical substances, but only a few, many of which occur in living organisms. Nor does the posterior formation of hypercyclusses between amino acids and proteins, the basic constituents of all living cells, seem to be governed by chance, but rather—as the leading German biophysicist *Manfred Eigen* surmises—by a constant adaptation of their functions to environmental conditions. Lastly, if chances for the emergence of life were really as small as *Hart* postulates, then—as the astronomer *Charles E. Seeger* pointed out sagaciously, "We wouldn't be here either."

After highlighting some of the most salient points of the pro and anti-ETI debate, which has been going on for some time, I think it is convenient to draw a bottom line and to pinpoint where the state-of-the-art regarding this intriguing subject stands now.

The skeptics have not been able to puncture the basic assumption on existence and frequency of life and intelligent life forms in other parts of our closer universe. The *Fermi Paradox* poses a valid question, but there are compelling economic reasons, which help to explain why even advanced extraterrestrial civilizations may not have struck upon our trail so far. On the other hand, man's search endeavors have been carried on for a period of time that is much too short to allow us to jump to the premature conclusion that no other intelligent species exist in our galaxy. *Tsiolkovki's* maxim, "Absence of evidence is not evidence of absence" still applies.

However, the question marks posed by the *Fermi Paradox* and recent scientific findings have forced the exuberant optimists to adopt a more cautious attitude and to review some of their exaggerated estimates. The majority of scientists engaged in SETI research today seem to agree with *Bracewell's* recent appraisal that "the Drake equation suffers from oversimplification." In particular, *Michael H. Hart's* findings, that the life-supporting orbit of a planet around its star is an extremely narrow band, strongly suggests that the estimated number of habitable planets in our galaxy must be revised. The euphemistic notion that our Milky Way galaxy is a busy playground of hundreds of thousands or even millions of space-faring and colonizing civilizations has rightfully come under serious questioning. No evidence warrants, however, the need to go to the other extreme, joining the gloomy speculations of *Bracewell, von Hoerner,* and the late *Josip Shklovskii,* recent renegades from the pro-ETI camp, who have warmed up to the idea that it may be up to the human race after all to preempt intelligent evolution in the galaxy and to colonize it from one end to the other.

As a result of this debate, a realistic school of thought, rejecting both the *Drake-Sagan* and the *Tipler-Hart* chauvinisms, has gained ground in the specialized scientific circles as of late. Important in this context is *Michael D. Papagiannis's* emphasis of the fact that it took only 0.7 to 1.0 billion years for life to evolve on Earth, but that since then, at least 3.5 billion years elapsed before intelligence and high technology developed, meaning "that the painfully slow evolution to high intelligence represents the bottleneck of the whole process."[16] By assigning to this time variable the very low value of about 10^{-6}, or only one planet in a million, *Papagiannis* arrives at about 1,000 advanced civilizations that have existed throughout the history of our galaxy. As the lifetime of an intelligent species, once it reaches technological and organizational maturity, is probably indefinite, the number of presently existing advanced extraterrestrial civilizations may be placed at several hundred, which—considering the task of achieving contact and maintaining a meaningful dialogue—is still a large number. In this case, the nearest intelligent extraterrestrial neighbor would live on a planet, distant well over a thousand light-years from our sun, which may explain why we haven't heard from him so far.

But even this much more conservative estimate is seen today as no more than a guess. Too many incognitos are still involved in the effort to quantify ETI in the galaxy in round numbers. The age of the universe, as it has been suggested recently, may be less than accepted so far, meaning that there would be fewer older stars than our sun. The factor of 10^{-6}, assumed by *Papagiannis* for intelligence to evolve on suitable planets, may be too pessimistic. On the other hand, longevity, even of advanced civilizations, may not be indefinite. For good reason, *Harlan J. Smith*, from the University of Texas, compared the search for extraterrestrial intelligence with a blind man in a dark room, looking for a black cat that might not even be there. As a result, a definite relativization of quantitative estimates of ETI has set in, even in pro-SETI quarters.

With regard to the fundamental issue—proving the existence of ETI—science is today placing the ball exclusively in the corner of empirical observation. There is wide agreement that the only way to know whether life is abundant or rare in the universe, and whether there are a million, ten thousand, or just a few hundred intelligent species in the galaxy, is through increased empirical knowledge of each of the factors of the improved *Drake* formula. The international SETI petition of 1983, signed by sixty-eight of the world's most renowned scientists, expressed this point very clearly, stressing that "the only significant test of the existence of extraterrestrial intelligence is an experimental one." With reference to the long and rather stale discussions on the *Fermi Paradox, Michael D. Papagiannis* recognized at the IAU Symposium in 1984, "We . . . began to realize that debates, useful as they might be in helping us to understand and focus on important issues, they were never going to solve this problem. The only way to solve it is the experimental search. . . ."[17]

In order to do justice to what has been achieved so far, it is also necessary to appraise the modest dimensions of SETI research, carried on until now, in a realistic way. With the fifty or so searches conducted so far in the radio, optical, and infrared part of the spectrum, at the very best a few thousands of stars have been covered, many of them with primitive methods and during too short a time, allowing no conclusive results. Much greater numbers of stars will have to be examined longer and more thoroughly before a solid estimate on the real frequency of intelligence in our galaxy or its absence may be reached. "The number of stars," NASA scientists emphasized, "that have been examined is less than 0.1 percent of the number that would have to be investigated if there were to be a reasonable statistical chance of discovering one extraterrestrial civilization."[18] This means that several million star systems will have to be closely scrutinized to arrive at a compelling conclusion.

With the aid of the new ambitious SETI programs presently underway in the U.S. and in other countries, many uncertainties still shrouding important variables, relative to the emergence of life and intelligent life forms, are likely to dissolve. With the help of the far-reaching NASA SETI program, scheduled to reach its climax at the turn of the century, the new telescope at the Allegheny Observatory, the space telescope, and equally ambitious Soviet and Japanese search projects, much deeper insights on this issue are bound to be achieved. In radio searches and in the search of planets around other stars, the new precision instruments, as *Papagiannis* stresses, "will permit us to detect even small earthlike planets in many of the nearby stars." The advanced state of science and technology, which our civilization is approaching rapidly, he went on, "will allow us to answer one of the most fundamental questions in nature—'How common is life, and especially advanced life, in the universe?' "[19] Mankind is likely to be in for some truly mind-numbing discoveries, maybe even before this century comes to a close.

General sentiment within the scientific community, engaged one way or another in SETI activities, remains therefore hopefully optimistic. It draws its conviction from the findings in a number of relevant research areas, most of which seem to point toward a universe endowed with a plurality of forms of life and intelligence. On this basis, the view of the skeptics, who pretend to know for sure that we alone are the pinnacle of creation, seem arbitrary and premature. It is difficult to brush off the suspicion that at the bottom of such opinions hides what *V.A. Firsoff* terms the "geocentric bias," the notion that "Earth is still the center of the universe, spiritually, if not physically," a prejudice ". . . constituting a constant source of error."[20]

"The history of science shows," Nobel laureate *Francis Crick* reminded his colleagues at the CETI conference in Yerevan, "that not only are the believers common, but that the skeptics are too skeptical." He went on, stressing that "it was only about 100 or 150 years ago that *August Comte* stated 'We shall never know of what the stars are made,' and that only forty-five years ago, *Lord Rutherford* affirmed that 'atomic energy was impossible'."[21] We may also recall that at the turn of the last century, famous physicists vouchsafed that we would never be able to fly, and that as recently as the thirties a distinguished scientist predicted that it would take man a million years to reach the moon. Consequently, it may be good advice to take visceral skeptics with a grain of salt.

In the final analysis, it doesn't matter much how many extraterrestrial civilizations may exist in our galaxy, as long as the probabilities are high that there are some. What really matters at this stage is to detect just one and to make contact with it. Detection would not only provide definite proof of the presence of life and intelligence elsewhere in the universe, but would mark the takeoff point of mankind for its fantastic journey to a new age.

One final reflection. What if, after analyzing a couple of millions of stars, our search endeavors by the year 2000 or 2020 fail to detect signs of intelligent life in the nearer regions of the galaxy? The proposition that intelligence is a much scarcer evolutionary end product and that, after all, we may be the only race capable of meditation, of distinguishing between good and bad, and in taking delight in the wonders of nature, would then gain in respectability. A feeling of cosmic loneliness would invade us, but not for long. Having enlarged enormously our knowledge of cosmology, we would draw renewed satisfaction and pride from the realization, that it may be up to us and to us alone, to set the galaxy ablaze with the torches of our intelligence. If this is the task that awaits man, it will no doubt sharpen his sense of responsibility for the conservation and grandeur of his species and its prodigious future here on Earth and in the stars.

Scientists investigating the border regions of theoretical and empirical knowledge are always treading on risky ground, leading sometimes to dead-end-alleys, but always within a step of unveiling one more mystery in which nature conceals ultimate truth. The search for extraterrestrial intelligence is one of these areas. In order to succeed, it will take hard facts, but also flights of bold vision, without which science would be a dull enterprise.

Chapter 15

ETI: Its Advanced Technological and Organizational Order

Perhaps our fears about extraterrestrial contact are merely a projection of our own backwardness, an expression of our guilty conscience about our past history.

—Carl Sagan

In what has been said on ETI so far, references about the probable existence of technologically advanced extraterrestrial civilizations abounded, but hardly any mention was made of species technically less developed than ours. One reason is that contact with a civilization possessing a far superior science and technology would, of course, be far more thrilling and rewarding for us than meeting a species of club-swinging cavemen. But the main reason is another one. In scientific circles, involved in SETI research, prevails the view that most extra-terrestrial civilizations existing in the galaxy are likely to be scien-tifically and technologically far ahead of us.

In this chapter, I would like to go a step further and substantiate the hypothesis that if there is intelligent life, dwarfing our most bril-liant technical achievements and scientific dreams, it must be equally advanced organizationally. Parallel with the development of a science and technology of great sophistication and perfection, such a superior intelligent race will necessarily have also raised its social and global political order to levels of extraordinary maturity and refinement. The significance of this proposition is not a trifling one. Contact with a superior science and technology would be of incalculable value to us. But what mankind may be able to learn from a much superior orga-nizational order would be of even greater meaning.

But let's take up first the claim that most extraterrestrial civili-zations are likely to be scientifically and technologically our peers. "The bulk of technical civilizations in the universe," wrote *Josip Shklovskii* and *Carl Sagan,* "may be immensely more advanced than ours—perhaps even billions of years beyond," and in *The Cosmic Con-nection,* the latter mused: "Civilizations hundreds of thousands or millions of years beyond us should have sciences and technologies so far beyond our present capabilities as to be indistinguishable from magic."[1] Other scientists, as I have shown, have presented convincing arguments, holding that there should be many advanced civilizations engaged in *Galactic Club* activities, building *Dyson spheres* around their home star, and sending starships or automatic probes in pursuit of expanded colonization and search of other life and intelligent species. The distinguished Soviet physicist *N.S. Kardashev* holds that super-civilizations of the Type Two and Three* are located not only on long-lived and slow-turning F,G, and K-type stars, similar to our sun, but in the center of our galaxy and in other galactic nuclei as well. Even *Freeman Dyson,* an ETI skeptic, has suggested that search efforts should concentrate on large stellar objects with strong infrared radiation, because "a civilization which exploits the total energy output of a star must radiate away a large fraction of this energy in the form of waste heat . . . emitted into space as infrared radiation."[2]

Just what the technological potentialities of a Type Two or Three civilization may be transcends the wildest flights of our fancy. As *Dyson* visualizes it, a Type Two civilization would be living in an artificial sphere, constructed at the appropriate distance all around its sun, making use of its total energy output. A Type Three would be a billion times more advanced. It would have colonized all of its galaxy, tabbed the energy of most or all of its stars, and built a truly galactic civilization.

Though the highly optimistic estimates of existing extraterrestrial civilizations—*Asimov* placed the number at 390 million in our galaxy alone, holding that almost all "are more advanced than we are"—have been whittled down lately, what I have mentioned shows that the conviction in the existence of technologically highly advanced civilizations is firmly entrenched in the mind of distinguished scientists.

What makes these scientists believe that? "Taking into account the fact," says *Kardashev,* "that the solar system is a second generation object, that its age is about five billion years, and that the oldest objects in the universe can be about twenty billion years, it becomes clear that the age of other civilizations . . . can be enormously greater than ours."[3] That most of the 200 to 300 billion stars in our galaxy are bound to be older than our sun, which was a rather late creation in one of its spiral arms, is—of course—a scientifically well-established fact, and the conjecture that on much older stellar systems, intelligence and

---

*Type Two civilizations have harnessed the total energy output of their star; Type Three civilizations, the energy of their entire galaxy.

technology must have reached levels infinitely superior to ours is therefore an absolute valid and widely accepted deduction.

Just what the science and technology of such a superior civilization may look like, is, of course, a matter of highly speculative nature. The following examples from the human analogy illustrate the difficulty. When asked about the state-of-the-art of laser technology, artificial intelligence, new energy sources or achievements of biogenetics in only three hundred years, the world's keenest futurologists are at a loss to venture more than vague guesses. If asked what our scientists and technologists will be able to do in a thousand or two thousand years—in cosmic terms, a mere trifle—they will rightfully decline an answer.

In much less than a thousand years, we will probably be mining some of our sister planets and their moons and circle the sun in *Dyson-like* spheres. Our thought processes and memory will be linked directly to supercomputers and data banks, and bioengineering will have bequeathed us with new and highly nutritious and useful plant and animal life. Industrial production and services will be run by fabulously intricate computerized and fully automatized robot systems, and our leisure time will by far exceed the daily working hours. We will probably have licked most of today's diseases and man's life expectancy will have reached over 150 years. But beyond that, except for a few risky extrapolations of a purely general nature, it is all wild guesswork. The attempt to glimpse the contours of a science and technology a hundred thousand or millions of years in advance of our present level, is clearly a task way beyond man's intellectual capacities.

Beings of such civilizations may be endowed with mental potentialities making us look like dumb mollusks in comparison. They may be capable of technical performances staggering our fanciest imagination. They may master speeds of thought ten thousand times faster than ours, as *Carl Sagan* speculated, and may have reached immortality, as *Frank Drake* has surmised recently. They even may be living partly in "free worlds," drifting through galactic space, and marshal energy resources, able to move planetary orbits with the same facility with which we butt a billiard ball. The practical value of such visionary gymnastics is evidently limited, but they help to illustrate that a civilization only a few thousands of years in our advance would make us look like bloody greenhorns in matters of science and technology.

The supposition that many of the extraterrestrial civilizations in our galaxy must be scientifically and technologically far more advanced than we are, is even more compelling if we consider the young age of our own technical civilization. Our industrial revolution stretches no further back than about 200 years. Our atomic, electronic, and space-faring era of rockets, laser beams, microchips, robots, hi-fis, huge radio-antennas, and biological factories became an overpowering reality in a span of not much more than five decades. Even by pursuing the roots of *Homo Tecnicus* a bit further to the empiricism and rationalism of *Bacon,* to the mathematics and astronomy of *Descartes* and *Galileo,* and even further back to the bubbling days of the early Renaissance and the curiosities of *Leonardo da Vinci,* the heliocentric theory of *Copernicus,* and the anatomist *Vasalius,* our technical age is not much older than half a millenium, a ridiculously short period on the time scale of cosmic evolution.

According to *Carl Sagan,* so far our civilization has lasted only for one-millionth the total lifetime of Earth; and this is including not only the Hellenic and Roman, but the earliest Babylonian and Syrian as well as the Sumero-Akkadian, Hindu and Sinic cultures, dating back to 3500 B.C. The following comparison is even more startling. Assuming for a moment that the lifetime of our civilization will be a hundred million years—a very conservative estimate—the 500 years of our technical age would be tantamount to only one two-hundred-thousandth of this period, less than an eye twinkle in a full long day. By equaling this life span of our civilization with a man's average lifetime of seventy years, then our present technological age is comparable with that of a human egg, just about three hours after it had been fertilized. By this yardstick, we haven't even been born yet as an authentic scientific and technological entity. What is more, we have just been procreated and are still in the womb, in a totally primitive and immature state.

A forceful conclusion imposes itself. If our scientific and techno-logical progress, attained during a ludicrously short period of time, has been so impressive, just what will the science and technology of a civilization look like that has reached adolescence or adult life with a technical age of millions of years! Our best scientists and engineers would probably view its wonders in frozen awe and with the same goggle-eyed expression with which the few remaining savages in the heart of the Amazon and Africa confront a TV set or a microelectronic chip.

It is therefore a legitimate proposition to suppose that even if an intelligent extraterrestrial species were just a few rungs ahead of us on the ladder of evolution, it would outdistance our science and tech-nology by a formidable margin. Considering that, in cosmic terms, our civilization has just arrived at the scene of technical prodigy, most older civilizations in the galaxy would outstrip our scientific and tech-nological standards by many magnitudes. From this perspective, con-tact with such a civilization would be a highly profitable undertaking, from which we would most probably have everything to gain and nothing to lose. Culture shock, a problem-issue to which I shall return later, may hamper the cosmic information flow for a brief spell of time. But in the long run, I believe the world's foremost physicists, chemists, biologists, mathematicians, astronomers, and computer scientists would have no qualms analyzing and trying to assimilate the extraterrestrial's advanced science and would pounce on it with the furious eagerness of a mining party that just struck upon a thick vein of gold.

But let's turn now to the second proposition, that most extraterrestrial civilizations are bound to be our peers not only in natural sciences and technology but also with regard to their global organizational order. So far, standard literature on ETI has dealt with this issue from a slightly different perspective. Are intelligent extraterrestrials likely to be benign or evil in the human sense of the terms? In case of contact, what would their reaction toward us earthlings be—friendly or hostile? The answer we give to these questions are of vital importance. If the aliens might be evil and potentially dangerous to us, obviously contact efforts should be avoided and no signals and sondes, giving away our location, should be sent into galactic space. If, on the contrary, probabilities are high that such beings would wish us no harm and would come to our encounter as friends and helpmates, then contact would benefit mankind and search of ETI should, by all means, be encouraged.

Popular science fiction still favors the first supposition. Since *H.G. Wells*'s book *The War of the Worlds*, a deluge of mediocre fiction, paperbacks, motion pictures, and TV thrillers feasts on the theme of the intelligent but wicked extraterrestrials, scheming to conquer Earth, to deprive us of our resources, to use us for their zoos or as a gourmet delicacy. The record-breaking picture *Star Wars*—its main plot a gruesome story of intrigue, power struggle, and wars of annihilation between a technologically advanced "good" mother planet and an equally advanced but "evil" rebel space colony—is an excellent example. The old classic *The Day the Earth Stood Still, Steven Spielberg's Close Encounters of the Third Kind,* and of course, *Spielberg's E.T.* are the exception rather than the rule. The great majority of Western authors of popular fiction keep almost pathologically fixed to the theme of the villainous extraterrestrial.

In scientific circles, on the contrary, the opinion prevails that intelligent extraterrestrials are likely to be benign. In his admirable book *Cosmos*, author *Carl Sagan* says that "it is pointless to worry about the possible malevolent intentions of an advanced civilization with whom we might make contact. It is more likely that the mere fact they have survived so long, means they have learned to live with themselves and others."[4] *Isaac Asimov* also believes "that civilizations that have managed to suppress undue violence on their home worlds would have learned the value of peace," and that "contact of minds across the great gaps of space would result only in good, not in evil."[5] Similar ideas have been espoused by the American anthropologist *Ashley Montagu,* and even *Gerald K. O'Neil,* though arguing that contact with an advanced extraterrestrial civilization may plunge us into a deep culture shock, concedes that "our galaxy is friendly and is waiting for them." In the Soviet Union, opinions on this matter are even more outspoken. "Contact between cosmic civilizations," assert *V. Sevastyanov, A. Ursul,* and *Y. Shkolenko,* "will prove to be peaceful and mutually advantageous, since the developed technology needed to establish such contacts is only possible where productive forces are very advanced and, consequently, social relations highly perfected."[6]

There is, of course, a remote possibility that we may be destined "to be engulfed by a greater civilization which is now expanding toward us," as *Bracewell* has suggested. But no evidence bolsters this supposition, and even if such an expanding thrust existed, it is unlikely that the approaching aliens are space pirates out on a treasure hunt. They would probably belong to a highly sophisticated species, interested only in increasing their scientific knowledge stock and with no desire of interfering in the evolution of another civlzation. In SETI circles, *Sir Martin Ryle's* exhortation against the Arecibo transmission in 1974 on the ground that, "By luring potential colonizers from far away in space we could bring about our own destruction, or suffer piracy and enslavement on a global scale," was rightfully fobbed off as a strictly minority opinion.[7] The same holds true with regard to the fear of the astronomer *Zdenek Kopal* "that we may find ourselves in (the extraterrestrial's) test tubes or other contraptions set up to investigate us, as we do insects or guinea pigs," and who pleads that if ever the *"spacephone'* should ring, "for God's sake, let us not answer."[8]

Attempting to draw a compromising line under this discussion, *Michael F. Papagiannis* admitted at the IAU Conference in Boston in 1984 that "After we debated for years until we were all red in the face, we finally began to realize that none of us could claim to know exactly how civilizations far more advanced than ours are likely to behave and act."[9] This frank statement would seem to corroborate *Freeman Dyson's* conclusion of twenty years ago that, "It is just as unscientific to impute to remote intelligence wisdom and serenity as it is to impute to them irrational and murderous instincts."[10]

Though in pure theory, this position, leaving open the possibility that highly developed aliens may be motivated by a code of ethics comparable to that of sharks, seems reasonable, I think it runs counter to the intrinsic logic of the evolution of technological intelligence. The crucial point, as I see it, is that speculations whether scientifically and technologically advanced civilizations are likely to be benign or not, is not enough. What needs to be examined closer is the likely global organizational, political, and social order of such civilizations. This scrutiny would seem to back up the contention presented at the beginning of this chapter and that I would like to introduce here as Proposition One:

> Scientifically and technologically far advanced civilizations are necessarily equally advanced organizationally and ethically.

The gist of this basic proposition is not entirely new. "It seems unlikely," wrote *Arthur C. Clarke* in the late sixties, "that any culture can advance for more than a few centuries at a time on a technological front alone. Morals and ethics must not lag behind science. . . . With superhuman knowledge there must go equally great compassion and tolerance."[11] Unfortunately this is still a rather unorthodox view. By and large, people, among them even a few scientists, readily accept the idea that extraterrestrials may be fantastically far ahead of our science and technology, but can't bear the notion that their societal order, too, may be quite different and much more advanced than ours, thinking that it would know greed and hatred, inequality, rivalry, violence, and war just as ours does. That this view is an illogical and scientifically unsustainable banality often escapes people's imagination. Civilizations with a highly sophisticated science and technology could not have reached this stage if they had not established along the way a peaceful and perfectly stable global organizational order. In its absence, they would have blown themselves to pieces eons ago, forfeiting all survival chances. To uphold the above notion means nothing less than negating that technological change necessarily engenders social and political change, changes of modes of production, ideas and values, and vice versa. It would mean sustaining the absurdity, that while the ETI's science, technology, and economy would have undergone cycles after cycles of revolutionary transformation and progress, its political and social structures and relationships, its ideology and philosophy, its value systems, life-styles and behavior codes would have remained the same, stagnant and unchanged for thousands and possibly millions of years. That this supposition contradicts all logic is self-evident. Our own short technical age is the best example. Evidence from the "human analogy" strongly supports Proposition One.

But first, what do I mean by an advanced organizational global order? In a nutshell, I mean a civilization, possessing a global peaceful order, a stabilized economy, and an egalitarian social system, free of antagonisms and violent upheavals, and with a planetary credo and culture of a high ethical content.

One aspect worth pointing out is man's adamant faith in progress. As *Robert Nisbet* has shown, by retracing the history of the idea of progress, the vision of humanity in continuous advancement, from a remote and primitive past to a distant but superior and glorious future, has been present from ancient times till today.[12] In spite of *Schopenhauer's* pessimism, *Spenglerian* prophecies of the inevitable decline of Western civilization and today's doomsdayism, this idea is still very much alive. Dynamism is still burning ebulliently in the furnaces of the leading capitalist societies, fostered by worldwide competition. Communism, too, baptized with the waters of Marxist and Leninist faith in human progress and revolutionary change, is pushing hard toward new frontiers. Third World nationalism, called by *Arnold Toynbee* "by far the most powerful of all living religions," is a progress and future-oriented driving force par excellence.

With regard to mankind's actual accomplishments, no one seriously dares to question the fabulous scientific and technical progress achieved during the past 200 years. But it is often ignored that the parallel social and political progress has been just as impressive. During these two centuries, in spite of intermittent retrocessions, breathtaking advances with regard to the scope of man's political freedom and his other human rights have taken place. Poverty is still far from eradicated, but the conspicuous rise of standards of living and working conditions of vast segments of the world's population cannot be dismissed lightly as a myth. Ruthless latifundist exploitation and the colonial yoke have been all but abolished and the long age of barbarian dictatorships finally seems to be coming to a close. The inhuman Nazi aberrational strand has been stamped out, and even in the Soviet Union, after dismantling the worst of *Stalinism,* Communism may shed some of its unsavory totalitarianism and acquire more democratic features.

The average person, overwhelmed by today's multifarious crises and everyday problems, is hardly aware of the tremendous social and political transformations that have swept the continents and remodeled the world in the course of only a few generations. It is those who lived the day when their country gained independence, who survived the horrors of the Nazi concentration camps; who, like *Anatoly Shcharansky,* made it from a Soviet prison in the Urals to freedom in Israel; who are keenly conscious of the fact that political progress is not a chimera but a reality, though it often appears a desperately slow and contradictory process. By taking the long and the large view, this particularity of man's recent history reveals itself with surprising sharpness.

Arthur C. Clarke's earlier quoted passage therefore needs modification. The suggestion that a culture may advance for a few centuries at a time on the technological front alone is not born out by our own evolutionary path. Only tribal societies without any, or hardly any, technological innovation over a longer period of time, vegetate also in social stagnation. Our civilization, on the contrary, having conquered in the astonishingly short period of only two hundred years the technical and atomic, the computer and robot, and finally, the bio and space age, has registered at the same time unparalleled social, political, and cultural progress. What is more, its most momentous thrust, the industrial revolution, now crossing with the exploration of artificial intelligence and applied bio-engineering into a still higher phase of scientific and technological progress, is unthinkable without the continuous social conquests and upheavals that followed it step by step, and that propelled it forward like an energy-producing dynamo. The French and the American Revolutions; the demise of feudalism and slavery; the retreat of traditional metaphysics, and the victorious march of the scientific world outlook; and last but not least, Russia's October Revolution and its reverberating impacts all around the globe to this very day, changed the social, political, and cultural make-up of our civilization as radically as the steam engine, the electric generator, atomic energy, and the microprocessor have revolutionized its technical and economic base. Steady advances in labor legislation, the great strides in the struggle against racism and on behalf of women's liberation, as well as the progress achieved in the fight for a healthier environment, provide further telling evidence of the ascending curve of mankind's social conquests. On the international scene, too, the emergence of the United Nations, the steadily increasing significance of Third World countries in the concert of nations and, lastly, the temporarily paralyzed but perdurable efforts to curb the armament race and banish the spectre of nuclear war, that at this writing again excites worldwide hopes, bear testimony to the fact that—considering longer periods of time—the belief in mankind's alleged social and political immutability is but a superficial misreading of history and our time.

It would be false, of course, to ignore the bitter social and political backwardness rampant in many countries and all that which still escapes man's endeavors for wider margins of freedom, participation, integrity, and well-being. Our evolutionary historic process has not proceeded along a steady linear curve. Periodic reversals and temporary stalemates making us taste defeat have been our continuous companions. In Part One of this book I tried to shed some light on the many organizational and ideological shortcomings still besetting our species, and on the hefty workload that awaits us in the future. But the general trend over the past 200 years and especially the past fifty years reveals an astonishingly consistent record of man's success to make life on Earth richer, more just, and more free and meaningful.

If this interpretation of our recent social and political development is correct, the "human analogy" would permit us to deduce that progress does not and cannot proceed along the one-track rail of science and technology alone, but that it moves on inseparably wedded to social and political advances, in a constant give-and-take relationship. A much older extraterrestrial civilization, possessing a highly superior science and technology, is therefore most likely to possess a peaceful and extremely perfected, smooth-running global organizational and social order.

Critical readers, though granting that over the centuries some social and political progress has been attained, still may interject that the final word has not yet been spoken and that *Homo sapiens* may still run amok and destroy civilization in a nuclear shootout. Though this possibility cannot be dismissed offhandedly, I don't feel so pessimistic about it and think—taking the long view—that before long, mankind will be approaching a stage in which the establishment of lasting peace and of a stable and more egalitarian one-world global order will become the feasible goal of Realpolitik. This optimistic appraisal of our future, which strengthens further the contention of the likely organizational superiority of ETI, is epitomized in Proposition Two:

> Though the organizational stage of our civilization is still immature, our world already goes pregnant with the seeds of a new political and economic international order, that will bring mankind in its peaceful fold.

At first sight, the supposition that our shaky civilization may be moving in the direction of such a superior organizational order, distinguished—as I visualize it—by the existence of a world state, the outlawry of organized violence, a globally run economy, national and social equity, the dominance of a cosmopolitan universal credo and a clear awareness of the species' cosmic dimension, may seem extravagant. On second thought, this may not be so. In an earlier chapter I called attention to the many unmistaken trends, pushing our world in this direction. The world state is still a utopia, but there is almost universal agreement among our civilization's foremost thinkers that it is the only solution of humankind, subjected to the abuses and arbitrariness of nation-states. Outlawing of war and collective violence still seems a far way off, but issues like conflict deescalation, reduction of medium-range nuclear stockpiles, larger-scale disarmament, and scrapping Star Wars all figure high on Soviet and American agendas for reaching a new agreement on detente and it is not impossible that at the end of the long tunnel of rapprochement the light of genuine peaceful coexistence might beacon. The exigency for the creation of a new and more just world economic order has stopped being just a clamor of Third World countries and is gaining ground also in the foremost industrialized nations, because of the growing awareness that the division of the world into rich and poor people, and into "have" and "have not" nations is unsustainable and profoundly disadvantageous for all. Nationalist and ideological chauvinism and religious sectarianism still keep driving dividing wedges between classes, sects, and nations, but there can be no doubt—*McLuhan's* "Global Village" is in the making. In the form of universal consumption, value and behavior patterns, the first offshoots of a global "world culture" are already thriving. "All nations and peoples," it says in a report to the Club of Rome, "are now members of an embracing global community. This is true whether we like it or not, and even whether we know it or not."[13] Innumerable bonds are knit this way across the boundaries of ideology, language, and color, fostering common credos and world outlooks. Man's cosmic dimension is also rapidly sharpening in focus. Such triumphant feats as the landing of the Apollo on the moon, and the explorations of Mars as well as the outer planets by the *Voyager* sondes are already celebrated as accomplishments of mankind as such, as apogees of the ingenuity and intrepidness of the human species. The likelihood that he may not be alone in the universe is also building up man's self-identification as a cosmic species and acceptance of *Constantin Tsiolkovski's* adage, "The earth is the cradle of mankind, but one does not live in the cradle forever."

I am not concerned here with the question of when all these worthy tendencies will fructify. It suffices me that they exist, indicating that something along the line of Proposition Two is in the making and maturing and this is what counts. In *The Third Wave*, author *Alvin Toffler* writes that "we are the final generation of an old civilization and the first generation of a new one."[14] This perspicacious statement may be prophetic in an even deeper sense. What seems to be evolving in a rudimentary, partly unconscious and almost demiurgic way, is not only *Toffler's* "Twenty-first Century Democracy" and "new culture," but the foundations of the take-off ramp, from which spaceship Earth will lift off to a superior global order. The scientific and technological revolution and man's rising pressures against nuclear war and poverty are bound to build the rest.

If this analysis is not entirely cross with what is really happening, it would corroborate the basic contention of this chapter even stronger. If our young technical civilization is likely to outgrow the dangerous stages of its present adolescent age, if it is already climbing up the tower, from whose springboard it will, in a not too far off future, take off to a new era of global and harmonious coexistence, it is almost certain that intelligent extraterrestrial civilizations, with a much older and more advanced science and technology, must possess forms of political and social organization of the highest order. Their organizational superiority should be of a comparable order of magnitude, distancing their scientific level from ours, meaning that their social and political order would probably be thousands or millions of times more developed than ours.

Charles L. Seeger is right in pointing out that this sort of argument on the basis of the "human analogy" suffers from an inbuilt flaw. "When speculating about the nature of extraterrestrial intelligence," he says, "we cannot escape domination by the anthropocentric predicament. We remain, as in the past, only able to devise totally fictional scenarios out of our earthly experiences and persuasions."[15] But if the "cosmological principle" stands, and if intelligence and technology are but culminating stages in the evolution of complex life, extrapolations from our history and experience are not tautological exercises but valid methodological aids. Apart from that, until first contact is achieved, they are the only ones we have at hand.

If far advanced extraterrestrials have rooted such destabilizing and disruptive factors as intraspecies rivalry, war, violence, ideological antagonisms and inequalities out of their existence, it is fair to assume that their ethical code, geared to conserve the peaceful and harmonious modes of their species' relationships in their habitat, would pose no danger to intelligent beings on other worlds. I would therefore like to propose Proposition Three:

> Advanced extraterrestrial civilizations would wish other intelligent species no harm. In case of contact, they would only have scientific interests and be willing to share their knowledge and experience especially with lesser developed intelligences.

There are probably other reasons also why such civilizations would not be a threat to us. The enormous galactic distances would necessarily act as shielding buffer zones, as would our ecosphere and microbiological life, which would pose considerable hazards to any non-artificial alien. It is also logical to suppose that even if a fugitive civilization were forced by its star or stars to abandon its habitual living space, and were looking for a new ecological niche to settle, it is extremely unlikely that it would choose faraway Earth, since its advanced technology would surely permit it to assemble from interstellar material a new agreeable habitat in much nearer star systems.

But I think the main reasons why we would have nothing to fear from a superior civilization is because, according to the golden rule, its ethical code would forbid any unfriendly and hostile action, not only because of scientific and altruistic considerations but because of self-interest. Contact even with less-developed intelligent races can, in the vastness of galactic space, full of unsuspected dangers, only enhance the survival value of their own species. We may therefore forget that the ETI's attitude toward us may be that of a ruthless conqueror or a doublecrossing competitor. If contact were achieved, it is much more likely that its beings would react with a serene scientific curiosity and that they would come to our encounter, as we should, with open arms, ready to teach and learn, and taking delight in the common bonds linking us across cosmic distances. I cannot imagine a different contact scenario, provided we are ready for it.

Two main conclusions may be drawn from the propositions I have tried to substiantate. One refers to our future. Our generation may not live to see the mentioned subcutaneous tendencies swell into a powerful mainstream and help create the new world order tailored to the needs of all mankind. But at least, as we approach the threshold of the twenty-first century, not everything looms bleak along the road. "There is no objective basis for the assumption," stresses *N.S. Kardashev,* "that some fatalistic law is operating in the universe to destroy every civilization several decades after it has entered the stage in which it can communicate, and that this law has already begun to manifest itself on Earth."[16]

The other is even more significant. If the probability is great that older extraterrestrial civilizations are not only scientifically and technologically, but also organizationally our superiors and that they are also likely to be benign, then contact with them would be enormously meaningful to us.

Chapter 16

Contact with ETI:
What It Would Mean to Us

*It is hard to imagine a more exciting astonomical discovery or
one that would have greater impact on human perceptions than
the detection of extraterrestrial intelligence.*
　　　　　　　　　　　　　　　—U.S. Astronomy Survey Committee

In 1958 the German engineer, Eugen Sänger, pleading for joint international space exploration as a means to abolish war, wrote that though
man may come across other life forms, "to muse about the cultural
consequences of a possible contact with extraterrestrial intelligencies
must be left at this time to philosophers and romanticists."[1] This was
only one year before *Cocconi's* and *Morrison's* far-reaching proposal to
search for ETI at the "waterhole" and only two years before *Frank
Drake's* "Ozma I" project, which established the search for extraterrestrial intelligence as a firm scientific discipline. *Sänger's* skepticism
about the timeliness of research regarding the possible consequences
of contact therefore couldn't have been further off the mark. The ink
of his book had hardly dried, when distinguished scientists and top-
ranking scientific institutions began giving the possible effects of contact with an advanced extraterrestrial civilization serious thought. For
instance, the NASA study, "On the Implications of Peaceful Space
Activities for Human Affairs," awarded to the Brookings Institution
and published in 1961, recommended, relative to ETI, research "to
determine emotional and intellectual understanding and
attitudes . . . regarding the possibility and consequences of discovering
intelligent extraterrestrial life."[2]

Since then, many scientists have looked into the possible effects
of contact with ETI, and some, like the Czech astronomer *Rudolf Pešek,*
have gone as far as suggesting that it even may provide the miraculous
cure to the problems besetting mankind today.

I think that contact with an intelligent extraterrestrial species
would be the biggest news ever, dwarfing in its significance such hall-
mark discoveries as that of the New World or of atomic energy. But

183

it will not furnish us overnight with magic formulas for getting rid of war and poverty. It will probably mark the take-off point to a new and uniquely thrilling and rewarding epoch of mankind, but it will not relieve us of the responsibility for solving our crisis ourselves. "The consequences of communication with extraterrestrial intelligence," emphasizes the U.S. sociologist *William S. Bainbridge,* "would be stunningly revolutionary," and *Carl Sagan* is convinced that "if we were to succeed, the history of our species and our planet would be changed forever."[3] But what does this mean concretely? To elucidate the issue, I think it is convenient to distinguish between the short and the long-term effects of contact.

With regard to the actual contact announcement, *Ian Ridpath* holds that, since in many people's minds extraterrestrials already exist (the UFO myth), eventual discovery of intelligent extraterrestrial life "would come as an anticlimax rather than a shock." But this seems unlikely. Discovery of another intelligent species would not only be "front-page news everywhere," as the report of the Brookings Institution surmised, but would certainly have a staggering impact on world opinion and stir worldwide floods of reactions. As its meaning would be assessed by leading scientists and commented on by hosts of publicists, media wizards, and certainly by all foremost statesmen, the significance of the discovery and its fantastic perspectives would surely ignite the imagination of people all around the globe. A wave of speculative but scientifically well-founded future scenarios regarding the possible advantages and benefits that Earth may derive from a technologically and organizationally superior species, would be touched off, and people everywhere would begin to realize, regardless of the time delay implicit in a cosmic dialogue, that after detection of another intelligence in the galaxy, nothing would remain the same on Earth. Confronted with the perspective of a revolutionizing information load, relative to a broad spectrum of scientific insight, it would dawn on us that man's history would have to be rewritten in terms of the evolutionary pre-contact stage and the truly revolutionary post-contact era.

One of the first consequences of physical contact or the much more likely radio contact would be, as the psychiatrist *Joseph Royce* pointed out, "an enormous de-encapsulating impact on man's world view, perhaps of the magnitude comparable with the Rennaissance and the subsequent Copernican Revolution."[4] *Carl Sagan* also foresees that in case of contact, "there will be a deparochialization of the way we view the cosmos and ourselves." The realization that we are not the only intelligent species in the universe and that other, more advanced beings exist, would necessarily hit man's egocentric hubris like a chilling shower. But, on one hand, this sobering experience would stir a giant wave of curiosity in the scientific, technological, societal, and biological make-up of the intelligent relative in space. On the other, it would spur a healthy relativization of man's disproportionate self-idolatry and help him sharpen his focus on the many unsolved earthly problems. Contact with a superior intelligence would thus trigger a bombshell-like explosion right in the ego of man, forcing him to live henceforth with the consciousness of the smallness of his petty earthly quarrels. Accustomed to think of himself as the pinnacle of creation, he would suddenly see himself stripped of some of his most cherished myths and forced to bear his sizing-down to what he really is—just one of probably many forms of self-regulatory and self-contemplative matter in the universe, and a less developed one at that.

Awareness that another intelligent race has reached high levels of technological and organizational excellence would also help the evolution of what *Curl Sagan* has termed the "cosmic conscience." Until today we live with that aching sense of loneliness *Loren Eiseley* spoke about in his book *The Immense Journey,* with that heavy conscience that "perhaps (we are) the only thinking animals in the entire sidereal universe." Proof of the existence of intelligent aliens would bust this sensation of forlornness and isolation in the vastness of cosmic space wide open and would help to instill in man, for the first time in history, the feeling of genuine biological and spiritual unity with the universe. Thereafter, man's gazing at the mysterious star-littered nightly sky would never be the same. Knowledge that another intelligence is roving out there, would demystify galactic space, with innumerable enigmas and dangers lurking in its impenetrable depth, and give it a friendlier and less awe-inspiring dimension. It would make man aware that he will be able to share his triumphs and failures with others, that bonds of cooperation and of great ratiocinative excellency and spiritual intensity may weld our intelligences together and that never again, in eons of the endless future, we will have to face the vagaries of cosmic contingencies and of destiny alone.

Last, contact with other intelligent beings will inevitably strengthen the bonds between man, fostering beyond all ethnic, political, ideological, and religious differences towards the idea of the oneness of the human race. The achievement of contact, maybe via reception of an artificial message, as a scientific feat of unparalleled significance, would necessarily cause a swell of universal elation and pride and would be hailed and celebrated not as a triumph of one nation alone but of all mankind. The need to beam a reply in the name of all mankind and the just expectation of all nations to share whatever beneficial information may reach us over the cosmic pipeline are bound to contribute to the molding of a global common and unifying world outlook.

But let us turn now to the second question. Once we have given the go-ahead signal to the celestial dialogue and have sent a codified message containing also an initial questionnaire in the direction of the ETI's home planet, what information, of value to us, may we expect? During the initial period, our receiving devices may remain quiet for a relatively long time. But once the two-way information exchange had intensified to a practically nonstop operation, it is very likely that mankind would receive, over a short period of time, an incalculable wealth of information, enabling man to leapfrog whole stages of scientific and technological evolution, and to gain insights in the structure and functioning in ETI's organizational order, existing now only in our dreamworld.

Due to the great galactic distances that would have to be traversed, the time that would elapse between reception of the first alien message and reception of their return message to our initial reply, would naturally be a source of no little frustration to the involved scientists and all people anxious to have a peep at the secrets and the appearance of the sidereal kindred. If our extraterrestrial interlocutor would have his radiotelescope installed somewhere in the *Constellation Ophiuchus*, about five hundred light-years away, and assuming that he would have no qualms answering our questions immediately, his return message would not begin trickling into our receiving antennas until a thousand years hence. In the meantime, generations of wondering astronomers will have passed away, and our present dire stalemate situation would by then be buried in the annals of history.

But, then, the contact scenario may present itself in a more favorable light. The extraterrestrial's home planet may be closer to our sun, which would slash the annoying time interval considerably. More likely, advanced extraterrestrials, aware of the hindrance posed to communication even at relativistic speeds, might beam modular information packages into space at regular intervals, with the hope and desire that other intelligent beings might pick up the signals and take advantage of them. Faster than light traveling message-carriers are another possibility that should not be ruled out. Finally there is the eventuality, however remote, of physical contact. Attracted either by our stellar messages or our expanding radioelectric wavefront, an alien spacecraft may suddenly be detected by NASA's distant sensing devices and express desire for landing. In this case, contact day may be any day and an exciting information exchange may get under way almost immediately, provided the visit takes place under the aegis of mutual trust and respect.

Trying to imagine the information load, which a transmitting advanced extraterrestrial civilization may dump into our lap, assuming the less propitious but more likely case of time-consuming radio communication, is a task I almost hesitate to tackle. Not because I fear that my sketchy account will appear too fantastic, but because I am afraid that it may be too conventional, too bound to earthly parameters and fail therefore to do justice to the fabulous surprises this civilization might have in store for us. In order to anticipate the breathtaking scope and revolutionary impact of the cosmic knowledge transfer, it is necessary to stretch our imagination, even if it means touching the borderline of the fantastic, and prepare mankind for it.

Such an information flow, says *Michael A.G. Michaud,* one of the few American political scientists interested in CETI, "could generate a dramatic forward leap in our sciences and our other academic disciplines," giving us for the first time the chance of comparing our knowledge with that of an alien intelligence, which may lead to "a boom in interdisciplinary sciences, and new branches of science."[5] Considering that our transmission rates already exceed 10^8 bits per second, it may easily be inferred what extraorinary avalanche of information of a superior civilization's science, technology, organizational order and history would hit our decoding desks.

Some of the received scientific information may puzzle us to no end and remain incomprehensible for some time. But it would be different with information sent in return to specific earthly requests. It is possible, writes *Carl Sagan,* that "we might receive . . . detailed instructions for the construction of material objects, a scale model of the capital of Delta Pavonis 3, a household appliance of Beta Hydri 4, or perhaps a novel scientific device of 81 Eridani 2." And it is also a possibility, as the British astronomer *Fred Hoyle* has suggested, "that we should receive detailed instructions for assembling the genetic material of an extraterrestrial organism, even an intelligent extraterrestrial organism."[6] Other scientists, like *Isaac Asimov* and *William Hamilton,* believe that from one message alone, "We may get new insights into still mysterious aspects of physics," or receive hints for "the cure of disease and (development) of better methods of growing food." Tapping the wealth of the galactic heritage, contends the NASA specialist *Bernard M. Oliver,* would make available to us "such things as pictures of the galaxy taken several billions of years go; it would include the natural histories of myriads of life forms that must exist in the planets of their member races . . ." and we would "learn their biochemistries, their variety of sense organs and their psychologies."[7]

Once regular two-way communication channels become operational, an unprecedented scientific rush to jump on the bandwagon can be anticipated. Research institutions from all over the world will vie and lobby for getting their inquiries and contributions onto the cosmic journey.

Theoretical, molecular, and atomic physicists would likely be most interested in information, enabling them to crack the "Unified Field Theory," vainly tackled by *Einstein,* to understand the true nature of the atomic structure and its subparticles and to construct a technically safe and economically efficient fusion-reactor. Molecular chemists and genetic biologists would want insights in the still largely obscure processes and phases of emergence of self-replicatory life from inorganic matter, in the codes ruling DNA and RNA functions and possibly, how to develop new useful plant and animal life or how to slow man's aging process. A brain specialist might wish to find out just how thought is produced, stored, and accessed, while virologists would probably be anxious to learn how to beat such implacable human enemies as the flu or AIDS. Geologists will be curious about exact methods for predicting earthquakes and volcanic eruptions and mining technologies in deep layers of the earth's crust and in our neighboring planets.

It is, of course, an illusion to expect from ETI ready-to-be-made formulas and turn-key blueprints to our specific needs. Because of its possibly quite different biochemical composition, material and ecological environment, and degree of automatization and abstraction, even an advanced extraterrestrial civilization may require time for understanding our physical reality and our life processes, before being able to beam useful information our way. Rather than proceeding along a linear question-answer pattern, our knowledge assimilation will probably be a process of creative reaction to the extraterrestrial's realities and scientific challenges. Our R & D scientists and engineers are likely to get more stimulating hints from analyzing the aliens' mathematics, physics, biochemistry, astronomy, engineering, and medical sciences than from the attempts to get ready-made handouts in the form of miraculous panaceas tailored to earthly needs. It is in this way that new revolutionary paradigms may be triggered in many scientific disciplines, and how technological spinoffs, ready for practical application, may become accessible to us.

But it is not only through the great epoch-making breakthroughs that hitherto unimaginable vistas will be opened up to our science. Even modest new insights might help us advance in areas of paramount importance. A few hints are all we need for controlling thermonuclear fusion and exploiting solar energy efficiently and on a large scale, securing abundant and cheap energy supplies practically forever. A few chemical formulas may put us on the trail of materials, light as feathers, but resistant to ages of wear and tear and a wide range of pressures and temperatures, while in another field we may learn how to make space colonization a spiritually rewarding and profitable enterprise. In microelectronics, knowledge of how to construct the ideal conductor in normal temperatures may help us jump to the tenth generation computer and robot, make full automatization of production plants a reality, and facilitate even direct link-ups of our brain with data banks.

In bio-engineering crops with nutritional values tenfold in comparison of those available to us today, double-sized cows, giving 150 quarts of milk per day and semi-intelligent house pets, taking care of groceries or doing the gardening, may well be the result of extraterrestrial information pouring into our biological laboratories. Much sooner than anticipated, we may leave our smoke and soot-spouting industrial behemoth complexes—monuments of ugliness, filth, and danger—behind and move on the entire front of chemical and pharmaceutical industries, to the environment-friendly factory, in which life microorganisms will be the producing agents of medicines and chemicals.

In medical research also, only a few penetrating ideas, unexplored so far, may put us on the right track in cancer research and in treating other diseases incurable till today. We might become acquainted with highly sophisticated surgery methods, with safe techniques of organ transplantation and limb regeneration and forms to adjust the inadequately synchronized growth rates of our neurons and hormones. The elixir for doubling or quintupling our life span may be placed within our reach, and if *Frank Drake's* prediction is correct, advanced extraterrestrials may even hand us the key to unlock the secret of immortal life.

Astronomy, space travel, and interstellar communication are bound to receive a powerful shot in the arm. One of the first concerns of the scientists at the other end of the cosmic dialogue would surely be the transmission of information for improving our sending and reception techniques, to make them compatible with their highest standard. There may be proposals for building antennas, capable of tuning in on their TV channels or for designing communication devices, based on faster-than-light traveling neutrinos or quarks as carriers, which would speed up the intragalactic dialogue and facilitate instant point-to-point communication between individuals on Earth. We would also learn whether *Einstein's* "time dilation" principle* would permit our

---

*According to Einstein's theory of special relativity, at speeds approaching the velocity of light, time passes and life processes take place at a much slower rate than on Earth.

voyages into the far corners of the galaxy, what propulsion systems we should use for our spaceships, how we may best terraform Mars or create an oxygen-rich atmosphere on Venus, while enriching at the same time our scanty knowledge regarding the creation of the universe. To the delight of CETI scientists we may even receive word of the existence of other intelligent species and be told how to contact them.

With regard to this brief excursion into the realm of the longer-term consequences and benefits of contact, one aspect deserves to be stressed particularly. Possibly with the only exception of immortality, none of the other hypothetical contributions to our natural, engineering, and medical sciences are propositions belonging to the province of sheer fantasy. I have referred myself almost exclusively to scientific propositions or projects that are either already on our drawing boards or that will be in our reach within a few generations. Actual knowledge transfer will necessarily take place on a much broader and more sophisticated scale.

From this it follows incontrovertibly that, once started and kept up, the bi-directional communication exchange with another advanced intelligence would give the whole spectrum of our science and technology a walloping kick forward. In comparison, the impact is bound to be immeasurably more dramatic than confronting *Copernicus* or *Newton* with *Einstein's* special theory of relativity or making *Lavoisier* familiar with *Francis Crick's* and *James D. Watson's* theory of the double helix. All basic sciences would be confronted with the necessity to subject their state-of-the-art to an agonizing but at the same time tremendously exciting reappraisal and to part with some of their most cherished premises and paradigms. More profound insights, hitherto concealed from our inquisitiveness, would replace them and probably change in time much of our present scientific world view. Stimulated by payloads of exotic and stunningly revolutionary ideas and insights that will make the hearts of scientists and researchers all over the world palpitate like wild seismographs, "We will embark," as *Bernard L. Oliver* put it, "on the greatest voyage of discovery in all our history."[8] Superior visions of creation, of celestial mechanics, of the origins of life and maybe even of our own distant past would substitute our present limited levels of understanding and new facets of the cosmos, of nature and of our potentialities may be revealed to us, permitting man to glimpse sights of unsuspected grandiosity and accomplishments. Access to an *Encyclopea Galactica,* as *Carl Sagan* envisions, would open our eyes to an informational wealth of unimaginable dimensions and value to us. But the biggest scientific breakthrough benefitting mankind will probably be those we expect least, that will fall into our lap almost inadvertently, emerging in the course of the tenacious sifting and assimilation process of the transmitted information, in which the cream of the world's natural scientists will be employed during ages.

But the long-term consequences of contact with a superior civilization will not be restricted only to revolutionizing the world's scientific and technological standards, but will also have a profound bearing on its organizational order, its political structures and ideas, and its value systems and culture. Since—as I have argued—the contacted civilization is likely to be highly superior on this score, its impact in this realm may turn out to be even of greater significance than the described benefits of our science.

It has been advocated often that the reception of a first message from the stars would have the meaning of a resounding gong for our crisis-stricken collective conscience. A signal of a far-off civilization, writes *Isaac Asimov*, would give us "the heartening news that at least one group of intelligent beings has reached our level of technology and has succeeded in not destroying itself, but has instead survived and advanced onward to greater heights." And he goes on, pointing out that contact "might even, perhaps, provide the crucial feather's weight that may swing the balance toward (our) survival and away from destruction."[9] Other authors also suggested that extraterrestrials might furnish us straightforward solutions for coping with such earthly problems as food shortages, population growth, and war, and for evolving into a peaceful and stable world society. Man looks desperately for answers to these basic problem-issues. No wonder that when the German astrophysicist *Reinhard Breuer* asked some of his colleagues what questions should be posed first to extraterrestrials, the most common answers were: How to solve the energy crisis and how our civilization could survive its present evolutionary stage, without destroying itself![10]

While such opinions certainly arouse much appeal, I think they urgently require a few qualifications. The crucial problems of our time as the East-West confrontation, the arms race, the North-South gap, and widespread poverty, are clearly issues mankind has to begin tackling here and now. We cannot afford crossing our arms and wait for the arrival of alleged panaceas from faraway stars which may not reach us for many decades and even centuries. By that time we should have licked today's most severe problems. Besides, it is worth recalling that it is not lack of rational alternatives which keeps stalemate in its contemporary deadlocked situation, but lack of global political consensus, that not even the wisest and most well-intentioned extraterrestrial recommendations would be able to propitiate from one day to another. *Reinhard Breuer* is therefore probably right, charging that the "belief of many people that the quasi-godly knowledge (of ETI) would save the world from all its ills . . . and that (via contact) we would find out how to conquer the present crisis, how to save the world from self-destruction, how to stabilize it and prevent genetic degeneration . . . is the fairytale of the galactic patriarchate."[11]

I am inclined to think that the extraterrestrial's great social and political contribution should be seen less in their capacity to provide Earth with ready-made recipes for its major ills, and more in the impact of their superior organizational order on man's consciousness and imagination. We should expect less in terms of detailed recommendations for a fairer international food distribution system or a tip-off on how to do away with the Soviet and American strategic nuclear missiles, and more in terms of the deep impression which, for instance, the absence of violence and inequalities in such an order would have on world opinion. To illustrate the point, let us assume that in relation with this subject we may be able to decode the following initial message:

> Ours is a world society governed by global law, binding for all members of natural and artificial origin. Our top executive functions, supervised by the Council of the Wise, are run by intelligent machines, which plan and control the productive and distributive system and the scientific and technological endeavors of our species. Though physically we still bear some distinctive traits, the remainder of a far distant past, our species is bound together by a common purpose, common values and a common culture. Our highest goals are to enhance our scientific knowledge, to maximize our society's growth and achievements, and to guarantee to its members an optimum of freedom, well-being and self-realization.

> Ours is a peaceful world—we know neither internal antagonisms nor external foe—with a globally planned, egalitarian and harmoniously functioning economic and social order. We know of neither poor nor wealthy individuals and nations, only of equals. And after having evolved along a long trail from primitivity to a complex and stable global order, we are moved by great compassion, watching other intelligent races struggling along a similar path, and are ready to tend them our aid and friendship.

I readily agree that this is a very humanized version of what some of the main characteristics of the extraterrestrial global organizational order may be. It might be a terribly trite and unimaginative image. Reality might be much more sophisticated and fascinating. But I think that even if we were to receive only this superficial description, its profound meaning would not be lost on mankind. In spite of its cryptical content, it would give social scientists and politicians a challenging box full of puzzles to solve and impress upon people's consciences all over the world that social and political maturity may be reached together with technological excellence, and that a peaceful order is an attainable goal for an intelligent species.

In subsequent phases of the information exchange we may learn a lot more about the intriguing facets of the extraterrestrial's organizational structures, their normative instruments, their centralized or highly decentralized administrative system, their ideological outlook and the ethical content of their intra- and interspecies' behavior code. Our statesmen and politicians would probably be in for a big surprise, learning about the highly automatized and therefore practically unconflictive consensus-reaching procedures, the lack of ruthless power politics and rhetorical demagoguery. Much to their dismay, our hapless economists would find out that a highly computerized and robotized and globally planned economy may perform miraculously well without the frustrating, costly, and unpredictable ups and downs that distinguish our world economy, and inbuilt shameless profit motives and dangerous springs of competition. The information that the extraterrestrial society had dumped the nation-state concept eons ago, and has been ruled ever since by a global constitutional framework that succeeds in keeping permanent peace and harmony, may spur naked envy on the part of all internationalists and be a shattering blow to all chauvinist nationalists, still hiding their fortunes behind the wall of the sovereignty dogma. It may also strike us as extremely unusual that its political and philosophical *Weltanschauung* and global culture would be astonishingly uniform in content and nature, and healthfully devoid of the belligerent pluralism and sterile traditionalism still dominant in many parts of our world. In the personal realm, last but not least, we may learn of relationships between their individual members of unfathomable intellectual and emotional intensity, to which we know no paragon, and capable of achievements of unmatched spiritual profoundness and aesthetic refinement.

We may now wrap up this discussion with the following important conclusion. Regardless of the informational wealth which our kindred in space may place at our disposal, and no matter of the spectacular scientific and technological advances which this information may spawn, the discussion of the long-term benefits of contact with ETI should not be linked to the expectancy and prediction of immediate stupendous political and social advantages for mankind. Contact with extraterrestrial intelligence does not and cannot relieve man of his obligation to find the remedies to his organizational maladies himself, nor can it help usher in the millennium. The finest alien blueprints for the creation of a peaceful and harmonious world order will remain an unattainable castle in the sky as long as our world, divided in antagonistic systems and rivaling ideologies, is not ready for it. But mankind will be able to reap one paramount harvest from continued exposure to information and images of an organizationally superior extraterrestrial civilization. It will sting man's conscience, stir his self-critical powers and thus far untapped fountains of imagination, and help spark currents of thought on behalf of profound changes of our unsatisfactory status quo. This alone makes CETI a tremendously worthwhile endeavor.

Skeptics, as I have mentioned in an earlier connection, hold that having reached a refined technical and organizational order, advanced extraterrestrials might find our penchant for violence, wanton materialism, and ideological fanaticism enormously distateful and repulsive. Learning of our leaders' grandiloquent proclamations of peace and justice and real political action, they—it is alleged—might want no part of us and would simply turn us the cold shoulder. The break-off of contact for this reason would surely be a humiliating, though not quite undeserved, lesson, but I don't think that the probabilities of such a "cosmic bum's rush" are very great. As *Ernst Fasan* speculated, some sort of *Megalaw* probably rules the relationships between intelligent extraterrestrial civilizations, and in this case they would abide by "the ethical principle that one race should help the other by its own activities," especially if the other is less developed.[12] But there is another reason that, in my opinion, vitiates the skeptics' pessimism. Though the level of our science and technology may not cause much excitement in ETI's scientific establishment, the fact that we are an intelligent species, our societal order and culture, our ecosphere and Earth's myriad of life forms would no doubt set their scientists' curiosity afire. Just as our researchers extract valuable knowledge from the study of primitive tribes and even primitive plants and microorganisms, so the extraterrestrial's R & D think tanks are sure to gain important new insights from our complex reality and life, which may help trigger unexpected spinoffs for their own use. Finally, there is also our civilization's rich cultural heritage, the range of our philosophical thought, the gold mines of our literature and fine arts, and the gamut of our music, which would certainly constitute the extraterrestrial's favorite pastures of investigation and study as well as for comparing their respective excellencies with ours.

Contact and two-way communication with another intelligent species will set in motion an adventure, pregnant with unprecedented potentialities for both participating parties. For our civilization, it would mean the starting signal for a journey in a fantastic future, propitiating for man for the first time a truly cosmic relationship.

But will communication and the exchange of valuable information with an extraterrestrial civilization be technically feasible at all? What if the two systems used for transmitting and receiving interstellar messages are different and incompatible with each other? The answer is simple—in this case, contact cannot occur. But on the whole, radioastronomers are optimistic. According to the Report on SETI, prepared for UNISPACE 1982 of the United Nations, decoding an extraterrestrial signal "would be relatively easy, because the signal should be anticryptographic, that is, made to reveal its own language coding. If the message comes by radio, both transmitting and receiving civilizations will have in common at least the details of radiophysics."[13] *Carl Sagan* also believes that "We will have little difficulty in understanding each other on the simpler aspects of astronomy, physics, chemistry, and perhaps mathematics."[14] As to more complex subject matters, *Philip Morrison* is probably right, pointing out that interpretations may turn out "to be a social task comparable to that of a very large discipline, or branch of learning," and that it may take time. Many radioastronomers expect to hit pay dirt with the new spectrum analyzers, that will be available by the end of this century, capable of scanning up to a billion frequencies simultaneously, and to begin decoding the first extraterrestrial message. If such a message were received either in the U.S. or the Soviet Union, we can take it for granted that neither would waste any time putting to work the most brilliant minds and the most advanced computers until its code is broken and its content understood. On the other hand, whatever communication method we may use, *Freudenthal's* language "Lincos," *Drake's* binary pictographs, or any other electronic way of transmitting content, charts, sound, and images, it is probably fair to assume that extraterrestrials, at least as versed as we are in radioastronomy, would have no trouble decoding our messages.

A few additional words still need to be addressed to the so-called "culture shock" theory. One of the first dramatic appraisals stems from *Carl Jung,* who felt that in case of contact with a superior civilization, "the reins would be torn from our hands and we would . . . find ourselves without dreams . . . and we would find our intellectual and spiritual aspirations so outmoded as to leave us completely paralyzed."[15] The same apprehension was voiced by the astronomer *Sebastian von Hoerner*, who warned, "If we come in contact with some superior civilization, this . . . would mean the end of our civilization."[16] Similar fears also must have haunted the physicist *Gerald O'Neill*, when he wrote, "The first effect of the discovery . . . would be to kill our science and our art. . . . Gone then is the possibility of new discovery, or surprise and, above all, of pride and accomplishment."[17] Other scientists, among them the U.S. physicist *Robert L. Sinsheimer*, have expressed identical caveats.

Parallels are invariably drawn with the destruction of the Aztec and Inca empires and the American Indian cultures in our own past, supposedly demonstrating irrefutably that collisions between superior and inferior cultures are disastrous for the latter. Though the examples from man's history speak for themselves, the parallel drawn with regard to the likely effects of contact with a more advanced extraterrestrial civilization seem, as I have tried to show, hardly warranted.

A more detached analysis reveals that analogies between the blood- and gold-thirsty Cortezes and Pizarros in Mexico and Peru, and the land-hungry settlers in the North American plains on one hand, and a benign civilization possibly hundreds of light-years distant on the other, are not convincing. Neither is it legitimate to suppose that ETI would be hellbent to intend anything like attempting to conquer and convert us with the "aid of the sword and the bible"—the password of the white man's takeover of the Americas—nor, as *Carl Sagan* has pointed out, are we going to be obliged to copy anything.

But the "culture shock" theory stands on shaky ground also for other reasons. Just because we may be confronted with a vastly superior science and technology, there is no natural law, forcing us into the jitters of a demoralizing "inferiority complex," or plunging our scientific estate into an emasculating paralysis. Quite on the contrary, stimulated by the prospect of accessing highly sophisticated and possibly revolutionizing scientific information, and the likelihood of achieving historical breakthrough on a number of fronts, and attracted by the lure of beaconing Nobel prizes and other awards, it seems much more likely that scientists of all nationalities and disciplines would pounce with unprecedented eagerness on the chance to get their hands on the extraterrestrials' know-how.

Furthermore, technology and culture-transfer flows should be regarded as indivisible facets of societal development. It is absence of these flows, not their abundance, that causes stagnation and intellectual attrition. Examples from our own history, such as the determined and highly successful introduction of the Industrial Revolution by nineteenth century feudal Japan, but also the present assimilation of modern science and technology and foreign political and culture patterns by many underdeveloped countries, shed an illustrative light on this fact. The U.S. did not fall prey to gloomy doomsdayism because of *Sputnik 1* nor is the European community reacting with decadent defeatism to the prodigious lead of the United States and Japan in microelectronics and bio-engineering. Why then assume that confrontation with a superior extraterrestrial science and organizational order would necessarily precipitate our world's scientific, intellectual, and cultural elite into fits of self-pity and hypochondria? Rather than making man's verve and creative ingenuity wither, I believe that contact with such a civilization would ignite in us the zeal to bridge the gap as soon as possible, to pick up the vitalizing glove of peaceful galactic emulation and to measure up politically and morally to the implied challenge.

Communication with a further advanced species will not be a one-directional affair in which we will play the subordinate role of receivers of extraterrestrial giveaways only. It will give humanity the unique opportunity to present to our relatives in space man's finest achievements. From the caveman to Leonardo da Vinci and Einstein, from the discovery of the New World to the touchdown on the moon, human history is an extraordinary odyssey with impressive testimony, speaking well of man's indomitable spirit and progress, against all odds. In our messages, beamed to the stellar interlocutors, we would certainly extol our noblest scientific and social achievements and present to them the magnificent record of our civilization's cultural and spiritual treasures, of which we can rightfully feel proud. Eventually, much of it will get interwoven in the broader fabric of a galactic culture, permitting another civilization and intelligent race to share our finest accomplishments and dreams. This knowledge should bolster our ego and our conviction that, as we approach a superior intelligent species, there is nothing to feel ashamed about.

Contact with a superior intelligence would therefore neither mean "the end of our civilization," nor "would it leave us without dreams," as it has been suggested. Rather than contracting an immobilizing "culture shock," man's innovative vigor, which has brought him so far, would receive an invigorating shot in the arm by its fabulous perspectives. Though it would sharpen his perception to the need of an overhauling change of our primitive world order, the mere glimpse of sublime forms of existence will not make him jettison what are real monuments to his greatness, and fall prey to bizarre extraterrestrial life-styles and exotic cultural tastes, no matter how magnetizing these may be. We may feel attracted by ETI's extravagant activities, its esoteric ideas and refined values, and we may even co-opt some of them, incorporating them into our culture and giving thereby our civilization a rejuvenating and maybe revolutionary impetus. The fact alone that from initial contact to the beginning of a meaningful dialogue much time will pass and that question-and-answer periods might well involve intervals, possibly stretching out lifetimes of generations, will surely act as protective shields to the alleged blows to our stamina and intellect and against cultural decay, which—as the skeptics hold—contact would trigger.

Far from entailing destructive or injurious consequences, contact and communication with an advanced extraterrestrial civilization, considering its far-reaching scientific and political offshoots, would herald a decisive turning point in mankind's history. It would help usher in a new age of accelerated organizational change and development of a geniune global conscience, and give man access to the cosmic stage.

It is probably an illusion to expect that the informational fortunes, hurled to us across space, could be translated in a landslide of immediate benefits. That will be the result of a slow apprenticeship that may require our most dedicated efforts for hundreds and maybe thousands of years. The most important repercussions of contact will be that it will turn man into a more astute observer and critic of the grave issues and challenges facing the world. It can help unleash the mental hurricanes needed to overcome our obsolescent global frameworks and hurry man's climb on the hazardous slope toward a higher order of earthly coexistence. This way it may contribute to bring this urgent task a step closer to fulfillment.

Once we know that other intelligent beings exist in galactic space, and that they have stretched out their hands toward us, nothing can impede us from reaching out and consummating the cosmic handshake. In the meantime, while the search for ETI goes on, we can only hope that its intrinsic message for our drifting civilization does not get lost and that it keeps shining, like a beacon of light in the dark.

Chapter 17

The Task Before Us

Our only hope is the energy-dispensing power, emanating from a new vision. . . . The utopian goal is more realistic than the realism of our contemporary politicians.
—Erich Fromm

Long-term consequences of contact with a superior extraterrestrial civilization will be of such colossal proportions for our world that we would probably be hard-pressed recognizing it. But this will be hundreds and maybe even thousands of years in our future. However, short-term consequences, as I have shown, will be of hardly less significance for us, and this—if contact is achieved soon—still during our lifetime. It would help to deparochialize man's world view, strengthen the bonds of unity in our species and, above all, contribute to an acuter and more demystified perception of the pitiful state of our world order. But this shattering impact on our oxidized political frameworks and fossilized ideological anachronisms will not take place if adequate search efforts are not carried out and if contact—once it is achieved—occurs in unpropitious conditions, in an atmospheric hold marked by prejudice, legal chaos, and confrontation, which may turn the historic event into a shameful fiasco. At this stage, mankind is still pathetically unprepared for the occasion, and I am afraid, that should contact happen now, its main function of giving man's complacent dreamworld a jolting blow, may be missed altogether.

In the first place, it may be seriously questioned whether mankind is psychologically and intellectually ripe for contact with extraterrestrial intelligence. Pseudoscientific beliefs that evil extraterrestrials will want to conquer our globe or that we will be saved by an act of grace by gallant UFO-driving knights are still very much in the vogue. *Von Dänikenist* elucubrations hold a strong grip on the imagination of many people. This mushrooming thicket of bias, together with the still abounding skepticism and the anti-scientific undercurrent celebrating a surprising resurrection, especially in the U.S., make it very likely that contact with ETI would be met today by Earth with a highly contradictory mélange of fear and exaggerated expectation and of outright hostility and rejection, rather than by a rational and friendly welcome.

The point in question is illustrated by our likely reaction to the physical appearance of the extraterrestrial aliens. *Carl Sagan* stressed that, "It is not a question of whether we are emotionally prepared in the long run to confront a message from the stars. It is whether we can develop a sense that beings with quite different evolutionary histories, beings who may look different from us, even 'monstrous,' may nevertheless be worthy of friendship and reverence, brotherhood, and trust."[1] As to the question if we are ready to sit down at a conference table and have a meaningful chat with an intelligent cockroach or squidlike creature, the answer is very likely no. But success or failure of the dialogue would depend on this readiness on our part and, of course, vice versa, because our looks may seem just as spine-chilling to them. Trying to imagine just what extraterrestrials might look like is, of course, a rather sterile exercise. As *Philip Morrison* has said, "They may be blue spheres with a dozen tentacles," or as *Walter Sullivan* has suggested, "as big as a mountain or as small as a mouse." Or—but this is a minority opinion—"The intelligent E.T. will basically look humanoid in form," as *N.J. Spall* maintains, arguing that extraterrestrials would have to possess organs for cognition and control as well as sensory and manipulative organs. But alluding to different planetary conditions and a different evolutionary path, most scientists are skeptical regarding the likelihood that an extraterrestrial species may resemble our human features, or look like the graceful angelic beings, or the big-headed but otherwise lovable E.T. in two of *Steven Spielberg's* outstanding pictures. The other possibility that a pictographic transmission may reveal weird or repulsively ugly beings, and that average people would react with instinctive aversion and hostility and want no part of contact, is more probable.

Similar doubts assault me, considering the likely intellectual and spiritual reaction of man to contact with an intelligent extraterrestrial species. Unquestionably NASA, the IAU Commission 51 on "Extraterrestrial Life" and many other scientific institutions in the U.S., the Soviet Union, and all over the world, would recommend strongly the prompt initiation of a constructive dialogue with our stellar likes. But how public opinion in Western nations, manipulated by media and influential special interest groups, and in many Third World countries, run by unpredictable leaders and regimes, would react, is another question. And it is still quite another matter, what governments as of the U.S. and the Soviet Union—the countries most likely to achieve contact—might decide to do.

While Communist countries would probably not hesitate authorizing a return radio message, the same cannot be taken for granted for the United States. Considering the vulgar conceptions warping public opinion on this issue and the lukewarm response that reception of an artificial message may elicit in conservative political quarters, it could well be that the necessary follow-up might hit the snag already on the initial basic question, whether the message should be answered at all and that the chance to establish a meaningful dialogue may get lost altogether in the ensuing torrent of inimical outbursts and official pusillanimity. Scientific minority groups, fearful that giving our whereabouts away to possibly evil ETI would entail irreparable damage to Earth and our species, may counsel extreme caution and to ignore the message. Misinformed congressman—let's recall *Senator Proxmire's* successful campaign against NASA's modest SETI project only a couple of years ago—may whip up public frenzy, painting gory doomsday scenarios and throwing wrenches in the Administration's decision-making process on the issue. Even more probable, fire-branding anti-science groups may raise hell and unleash nationwide outcries of protest, heaping disparagement on the feat, and mobilizing their most vociferous influence peddlers for preventing the transmission of a reply message. Fervent gurus, otherwise prone to preaching peace and brotherly love, might hurriedly orchestrate a strident hubbub and demand, "to kill and destroy the bastards" in case of an approaching extraterrestrial spaceship.

But there is still another huge roadblock which, under present conditions, may render inoperative a fruitful information exchange with advanced extraterrestrials. Harvard biologist and Nobel laureate *George Wald* alluded to it, stressing that contact with such a civilization would produce "the most highly classified and exploited military information in the history of the Earth."[2] In view of the deep-seated rivalries between the two superpowers, the appropriation of even a minor aspect of the extraterrestrial's probably much superior science and technological know-how would present either the U.S. or the Soviet Union with a windfall breakthrough, giving it very possibly a decisive military and strategic superiority. That such calculations—though they are not publicly advertised—are very much on the mind of American and Soviet military planners and strategists, can be accepted as a foregone conclusion.

Let us suppose that the U.S. may obtain from visiting extraterrestrials or via a radio message information on a propulsion system, reaching 0.3 c (or a third of the speed of light), which in case of contact is not at all an impossibility. Adapted to military need, this information may permit the U.S. to produce missiles with a speed of about 90,000 kilometers per second. They would be about three thousand times faster than those available today, about equal to the difference between a camel's trudge and the speed of a jet. Alerted of a pending Soviet attack, the U.S. could with such fast-flying missile-killing warheads not only pick the Soviet ICBMs and SLBMs off the sky like sitting ducks long before they reached U.S. targets, but destroy them before they left their silos. The dream of SDI would become a reality. ETI's technology would render the U.S. space invulnerable to a Soviet surprise attack, give the U.S. a tremendous offensive advantage, and tilt the existing balance of power irremediably in its favor. Or, in the opposite case, in favor of the Soviet Union.

This one example—many others are conceivable—illustrates the point. In case of a surprise landing of an extraterrestrial craft, Soviet and American special forces would attempt to hijack it, if that were feasible, in order to subject it together with its crew to the most meticulous scientific investigation ever. Military exigencies would very likely take preference over the humanitarian considerations. The same is likely to occur in case of radio contact. Here, too, military priorities may impose themselves and dictate reply terms. The danger is therefore not farfetched that contact with a superior civilization—once it is verified—may quickly be co-opted by the military high brass and R & D tanks of either of the two superpowers and that, consequently, the potential dialogue and knowledge acquisition may be very much tailored to their sinister ends.

Still another conceivable barrier for contact's role as an agent for a rousing great human cause may arise from a government's decision to cast a mantle of top secrecy on the discovery. *Charles Berlitz* exploited this possibility in his thriller *The Roswell Incident,* in which he rehashes the fantastic story that in 1947, an extraterrestrial craft crashed in New Mexico and that the debris of the wreckage and bodies of its crew were whisked away by the U.S. Air Force in a hush-hush operation. He alleges that the CIA was put in charge as a watchdog to keep this top secret from leaking out.

Whether firm evidence of existence of extraterrestrial intelligence in our galaxy could be kept under tight security wraps is debatable. In the U.S., the news of an authentic UFO landing or of an artificial radio message could probably be kept secret no longer than it would take the nearest TV station to dispatch its reporters to the scene. *Arthur C. Clarke* once quipped that keeping an extraterrestrial craft a secret would be "as difficult as hiding an elephant in Manhattan." But an authoritarian regime, as that of the Soviet Union, may want to clamp down on the breakthrough while trying to exploit it to the hilt for its own selfish purposes.

But even if this would not occur, enough hurdles would still have to be cleared. Because of the lack of adequate international legislation on the subject—in case of an approaching extraterrestrial spacecraft—the mere fixation of the tentative touchdown site may touch off a grueling haggling match, forcing the visitors to head back in total frustration. To reach a consensus on the content of a reply-message would also be an almost superhuman task, considering the deep divisions still marking international relations. Many psychological, pseudo-intellectual, legal, and political obstacles therefore still exist today that may impede a quick and rational response of Earth to contact with ETI.

Since irrefutable evidence proving the existence of other intelligent species in the galaxy may be in our hands sooner than the skeptics hold, helping to overcome these hurdles is not a merely academic question, but a matter of considerable practical urgency. What can and should be done?

Efforts in this direction, I think, need to be approached, recalling that search of extraterrestrial intelligence is an undertaking concerning man as such, and that contact with it must bring advantages and benefits not to one nation alone but to humanity as a whole. In this spirit, the participants to the earlier mentioned Conference on CETI in Yerevan in 1972 had demanded that, "The search for extraterrestrial intelligence should be made by representatives of the whole of mankind." The authors of the NASA SETI study of 1977 went even further, postulating that, "The SETI effort should be cast as a cooperative international endeavor at the start and that appropriate international relationships should be established through existing or novel international arrangements."[3] But one conclusion can be reached.

The only entity suited to promote SETI and CETI in the name and for the benefit of all mankind is the United Nations. Though the UN system, as shown earlier, is fraught with more than a few shortcomings, it is the only global organization representing some 150 nations of the world. *Ashley Montagu* once suggested that in case of contact, "independent bodies be set up outside governmental auspices, outside the United Nations, operating possibly within or in association with a university, whose object shall be to design possible means of establishing frank and friendly communicative relations with beyond-Earthers."[4] But because of the universal significance of the issue at stake, I see no way of sidestepping the United Nations. Only the UN would be able to speak genuinely for humankind as a whole. Only the UN provides the organizational setting, in which also Third World countries can press for the proper humanitarian orientation of search projects and contact modes. Along this line of thought, the CETI Committee of the IAF/IAA has urged that the Committee should "incite the United Nations' interest in CETI and help them in all CETI problems," as well as "endeavor that all important decisions concerning CETI should be taken by the United Nations."[5] But, unfortunately, the issue of ETI has received at the UN only lukewarm response so far.

The UN Committee on the Peaceful Use of Outer Space raised the question at its twentieth session in Vienna in June 1977, but in the end, SETI was assigned to the Scientific and Technical Sub-Committee, where it still reposes peacefully and undisturbed. However, in the aftermath of *Kurt Waldheim's* message to extraterrestrials in the *Voyager's* plaques, it posed to the UN Secretariat the following meaningful questions:

> What should be the content of future messages? Would work of art represent humanity better than samples of scientific knowledge which extraterrestrial civilizations can be assumed to possess? Should messages be composed under the sponsorship of an international body such as the United Nations, if they are to represent mankind? Should a register of messages to extraterrestrial civilizations be established and maintained to facilitate further references, since a reply to a given message may take a long time to reach the earth? Should the preparation of messages be done directly at the United Nations as an activity which would unify mankind?[6]

These pertinent questions remained unanswered. No proposal or draft of an international treaty regarding contact with ETI or joint search efforts was elaborated. No initiatives to fill the existing legal void and to establish some binding rules with regard to the important issues, that would have to be dealt with in case of contact, were put forth.

Lodging the required competence in the United Nations should not be an insurmountable task. The Space Treaty of 1967, which regulates the peaceful exploration and use of outer space, could serve as a useful stepping stone. It has consecrated this valuable principle:

> The exploration and use of outer space, including the moon and other celestial bodies, shall be carried out for the benefit and in the interest of all countries, irrespective of their degree of economic or scientific development, and shall be the province of all mankind.

This treaty, signed by the U.S. and the Soviet Union, forbids the national appropriation of the moon "by claims of sovereignty," and "by means or use of occupation or by any other means," and it also establishes that "all stations, installations, equipment, and space vehicles on the moon and other celestial bodies shall be open to representatives of other States party to the Treaty on the basis of reciprocity," and that "astronauts shall be regarded as envoys of mankind in outer space."*

These clauses of the Space Treaty clearly encapsulate the "international principle" that should also rule the endeavors of mankind in the realm of SETI and CETI. If the exploration of the moon and space in general is regarded, per se, as an undertaking that must benefit all mankind, the same principle must apply *a fortiori* to search of extraterrestrial intelligence and contact with it, which promises to report to our species benefits of incommensurably greater significance than those probably imparted by exploring and even colonizing the moon.

But no serious initiatives have been forthcoming so far, pressing for an early international agreement on the subject. Because of the general crisis affecting the whole UN system, but mainly because of the continuing virulent East-West antagonism, neither the U.S. nor the Soviet Union saw fit to give SETI and CETI the high-ranking priority for placing it on the top agenda of the United Nations. Another important reason is the conspicuous absence of the Third World in the intriguing debate on extraterrestrial intelligence. With the only exception of the clownish attempts of *Eric Mathew Gairy,* former Prime Minister of Grenada, to get UN attention and legitimation for the UFO myth, developing countries have played a surprisingly passive spectator role on this issue. Evidently, in the ruling elites of two-thirds of mankind, overwhelmed by the general crisis syndrome, the eye-opening process with regard to the potentialities of search and contact with ETI is still in an immature stage, though so much is at stake for the poor nations. Facing the threat to become disconnected from the train of human progress as a result of the burgeoning microelectronic revolution in the industrialized North, an even more disastrous future scenario of getting uncoupled is shaping up on the horizon regarding extraterrestrial intelligence—the danger to be left out completely from the most thrilling and profitable adventure of mankind and to get off it emptyhanded.

*The same principle, upholding the international character of space and space exploration, distinguishes the Agreement on the rescue of astronauts, the return of astronauts, and the return of objects launched into outer space of 1968, as well as the convention on international liability for damage caused by space objects of 1972.

To avoid this perspective, what should Third World nations do? I think it is precisely with the United Nations, where they might bring to bear their influence with the following aims:

1. to rescue SETI from the undertow of the East-West confrontation and to strengthen its humanitarian dimensions;
2. to promote the idea that SETI and CETI are enterprises of all mankind and that in their efforts and eventual benefits, all countries and humanity as a whole must participate;
3. to boost political consciousness on ETI and help create the legal provisions which will transform search and contact of ETI into genuine international endeavors.

The cause of the Third World would gain from such initiatives not only for strictly scientific and technological reasons, but because of their great potential for unifying people and making them more conscious of the great gulches dividing the globe and, above all, because it is necessary to ensure that the cosmic knowledge transfer, after contact with an advanced extraterrestrial civilization is achieved, will be used with the objective to do away with the terrible scourge of underdevelopment.

But which are the concrete steps that should be considered at the level of the United Nations for transforming SETI and CETI into effective and truly international endeavors? The most urgent need, I believe, is the creation of a *UN Committee for Search and Contact with Extraterrestrial Intelligence* with the following main objectives:

1. to prepare the draft of an international agreement, proclaiming the universal character of all research and other activities relative to search and contact with ETI, and the principle that they must be carried out in the interest and for the benefit of all mankind;
2. to serve as a clearinghouse of all SETI and CETI activities, and to promote the basic propositions on ETI as a strong world-uniting idea and undertaking;
3. to propose the set-up of an international *ETI Search and Contact Center* for research and rendezvous purposes;
4. to coordinate SETI strategies and establish rules and procedures for likely contact scenarios;
5. to establish ground rules for the transmission of internationally agreed upon message content directed to ETI, and for reception and decoding of interstellar messages;
6. to promote universal efforts with the aim of preparing man for the existence of other intelligent life in the galaxy, and making him conscious of the potential benefits of contact;

7. to prepare an international agreement for the case of an approaching extraterrestrial spacecraft, and to work out operational plans for mankind's response, if the intentions of the aliens are friendly, but also for the eventuality that they might be hostile.

With the creation of this *UN Committee for Search and Contact with ETI,* an important void in international law would be filled. If an artificial radio message were received today, and the necessity of a reply would arise, in the absence of international regulations, it is likely that a terrific scramble would break loose and that in the ensuing confusion and counterproductive scientific and political competition between the nation that received the signal and all other countries, chances for an equanimous reaction and the initiation of a meaningful dialogue may be condemned to failure. If an alien spaceship would suddenly appear on our radars, no international provisions exist for facing up to the emergency according to a predetermined action plan. A hostile attitude on the alien's part, which we can't afford to rule out completely, would catch us totally unprepared. In either case, unpreparedness and ad hoc improvisations might jeopardize the required serene and intelligent joint management of the momentous occasion.

The visit from an extraterrestrial spacecraft may not be the most likely scenario in the near future, but the least we should do is to create adequate conditions in order to be able to take advantage of the eventuality. The proposed *ETI Search and Contact Center* would serve these purposes:

1. as the ideal landing site, avoiding a contest between Cape Canaveral and Baikanur;

2. as a protective shelter, providing isolation and immunization of the extraterrestrials and their craft, to prevent potentially harmful microbiological contamination;

3. as research facility, permitting examination of the foreign craft, its material, instrumentarium, and propulsion system as well as the physical, chemical, and physiological make up of the aliens;

4. as a center for search activities in the whole spectrum of visible light, radio waves, X rays, ultraviolet and red light, equipped with adequate sending and receiving facilities;

5. as a communication center with the aim of developing a variety of communication methods with other intelligent beings;

6. as a ward, providing adequate defensive facilities, should the need for destroying the extraterrestrial expedition arise.

This *ETI Search and Contact Center* should be staffed with a scientific task-force representative of mankind, and work out propositions and operational plans for the diversity of imaginable contact scenarios and contingency situations that may arise in connection with the need for urgent reaction and response.

The creation of such an international ETI Center—quite in line with today's trend toward science-cities—bringing the cream of ETI specialists together, and benefitting from the large expertise available in scientific institutes of the U.S., the Soviet Union, and other countries; and that could still be largely enhanced by cross-disciplinary studies in many related fields, could no doubt give a tremendous boost to today's highly dispersed and uncoordinated SETI efforts. On the other hand, it is quite conceivable that direct interpersonal and interstellar communications with intelligent extraterrestrials might fail because of language deficiencies on our part. To forestall this possibility, a much wider range of sophisticated communication methods than *Freudenthal's* "lingua cosmica" and *Frank Drake's* binary code pictographs could be developed at this center.

With regard to interstellar contact, the British scientist *Duncan Lunan* analyzed in detail how it may take place, stressing that in either situation, the nuances between initial distrust, caution, a friendly or a hostile attitude on either side may be hair-thin. Contact procedures, especially in the case of physical contact on Earth or in space will necessarily require a high degree of psychological and political expertise that cannot be left to chance or allowed to be summoned up at the spur of the moment. *Lunan* proposes that "there should be a set of contingency plans to bring a suitable international organization into being at a few hours' notice."[7] But this proposition, even if it were feasible, would probably lead only to lukewarm and short-sleeved decisions. The legal provisions and procedural plans for a gamut of possible contact scenarios would have to exist beforehand—a very important job also for the center's political scientists, psychiatrists, and lawyers.

Last but not least, protection of Earth's and our species' vital interests is of paramount importance. Though everything speaks for the likely benign character of advanced extraterrestrials, precautionary strategies for every imaginable contact scenario must be worked out, ready to be put into action without delay. How should the commander of an earthly space station react to suspicious maneuvers of an alien spaceship appearing suddenly on his radar screen? What should our response be, if spider-legged creatures stalk out of a just-landed cylinder-shaped craft and start collecting samples of our flora and fauna, without paying the least attention to our signals inviting them to an

amiable roundtable chat? Considering the importance of neither mis-
interpreting the visitor's intentions nor wanting to arouse unwarranted apprehension on their part, the step-by-step approach, in the
case of danger—a warning signal, an attempt to scare them away, or
the decision to push any of several red buttons for incapacitation of
the craft or total destruction—should all be carefully planned by the
center well in advance.

There are, of course, many other scientific tasks that may be tackled by this *ETI Search and Contact Center,* but its main significance
is that it would launch SETI and CETI as genuine international and
planetary enterprises in which all nations would have a stake.

The need to invest the UN with basic authority regarding search
and contact activities of ETI by no means implies a reduction of the
respective efforts on the national level. Quite the contrary. *Michael
Michaud* is correct when pointing out, "We have reached our present
stage of extraterrestrial activity more through international competition than through cooperation of conscious species strategy."[8] The
United Nations continues to suffer from a severe credibility crisis, and
it will take time for Third World countries to begin flexing their mus-
cles for hierarchizing the ETI issue. Though its internationalization
should be furthered with all the efforts we can marshal, for some time
to come it is up to the national science estates, especially in the U.S.
and the Soviet Union but also in other space-oriented countries, to
push on an even broader front and more intensively the pertinent
research endeavors in this field. In comparison with the huge outlays
for space exploration, the resources allocated to SETI are a petty knick-
knack. Considering the incalculable benefits, which may be accessed
through contact, it is difficult to imagine a worthier investment of
industrialized nations than enhancing their funding of search activities
of extraterrestrial intelligence by several factors. For the United
States, this demand admits no discussion. SETI is one race which the
U.S. can not afford to lose. Quite apart from the possible spin-offs, a
Soviet first may well deal the U.S. prestige a humiliating propaganda
defeat from which it may have a very hard time recovering. But stepping up its commitment and efforts on behalf of SETI and facilitating
Third World participation in these endeavors would also be a politically
profitable undertaking. For reasons that are hard to explain, the Western world, and particularly the U.S., are singularly inept in captivating
the imagination of the broad masses of the poor developing countries.
By sponsoring Third World SETI activities and taking a lead for the
described internationalization of the issue and major undertakings at
the UN level, the U.S. may be able to touch off a surprising landslide

of sympathy in these countries and exert a lasting appeal to people everywhere on behalf of a cause that belongs to all of mankind.

Though competition between the two superpowers will continue to provide powerful stimuli for ETI-oriented research, Soviet-American collaboration in this exciting new field of science stands out as a perspective worth exploring. Now, in the aftermath of the Geneva and Reykjavik summit talks between *President Reagan* and *Chairman Gorbachev,* which have given rise to a ray of hope that the U.S. and the Soviet Union may be sensible enough to call off the deadly game of nuclear brinkmanship and are ready to accommodate themselves to a more stable relationship, the time could be ripe for a modest beginning. Political and ideological differences need not be insurmountable barriers to constructive collaborations in areas with a common denominator of interest, as was shown by the Agreement between the Soviet Union and the U.S. on Cooperation in the Exploration and Use of Outer Space for Peaceful Purposes, signed by *President Nixon* in Moscow in 1972. This agreement served as the legal basis for a series of highly significant joint scientific projects, among them the data exchange between scientists of both countries during the Mars exploration by *Mariner-9* and the Soviet probes *Mars-2* and *Mars-3,* the "Bering Project," and joint biomedical research pertaining to space flights. It also propitiated the complex Soviet-American cooperation for preparing compatible rendezvous and docking systems for Soviet and American manned spacecraft and stations that led in 1975 to the much publicized flights of the Soviet *Soyuz* spacecraft and an American *Apollo* space vehicle and their historical rendezvous maneuvers. Joint efforts for searching extraterrestrial intelligence and preparing contact could be initiated and carried out in the same spirit.

Since the low of mutual relations between the U.S. and the Soviet Union during the early eighties, circumspect attempts on both sides to avoid tilting the scales toward all-out war and to bring about a new thaw, have led to the reactivation of the Joint Commercial Commission, new rounds of disarmament talks and also to a slow but gradual increase of scientific contacts between the two countries. At the tenth anniversary of the *Soyuz/Apollo* rendezvous in July 1985, Soviet and American scientists considered possibilities of reinitiating joint space ventures. One possibility, entertained by NASA as well as the Soviet Institute of Space Research, is "a joint U.S.-Soviet manned landing on Mars."[9] If this promising trend continues, apart from this project, I can think of no other more meaningful area to start a collaborative effort, than to cap the pioneering undertakings of the seventies with an agreement on a joint ambitious SETI and CETI program. Such an undertaking would not only help to recreate currents of mutual trust and friendship, but if decisive breakthroughs were to be achieved under the umbrella of such joint efforts, its conflict-deescalating effects may well usher in a new era of global coexistence.

However, even colossal scientific achievements do not exert a shattering impact on people's consciousness over night. Attitudinal and behavioral changes—as development work in Third World countries reveals time and again—occur at a snail's pace and they hardly occur at all if general political and cultural patterns are not changing simultaneously. To wait for a great scientific breakthrough to give man's tradition-based outlooks an electrifying shock is not enough. The significance of what search of extraterrestrial intelligence and contact with it means for mankind must be stripped of its elitist and esoteric character, permeate public opinion and become interwoven into the matrix of man's cultures and value systems. Much can and needs to be done in this respect. To transform SETI and CETI into great conscience-stirring human causes, capable of mobilizing innovative currents of thought and the needed social pressure for change, it is necessary to enlist support from many quarters. Among these, the following play a particularly decisive role.

Social Sciences. With a few worthy exceptions, as I pointed out earlier, search and contact with ETI have so far functioned very much like a "closed door affair" of natural and engineering sciences. Astronomers, physicists, and molecular biologists have set the blazing trail and are still very much in command of this fascinating new scientific realm. By and large, social sciences continue to be conspicuous absentees in what already is the most exciting scientific endeavor of all times.

Should contact occur in the near future, social sciences would stand appallingly aloof from an event which will shake man's conscience more than any other scientific discovery or social upheaval in the past. They would have to jump on the bandwagon of CETI as latecomers by default and almost emptyhanded. This situation requires an urgent change. The possibility of contacting an advanced extraterrestrial civilization encapsulates a challenge of such monumental scientific and technological, but also political, social, and cultural implications, that it cannot be the exclusive domain of any branch of sciences. Sociologists, economists, political scientists, and futurologists, but also psychologists, philosophers, and even anthropologists may find in these implications a tremendously rich and almost unexplored terrain for important future-oriented research and for presenting our bereft species a new global design. They could not only prop up the fundamental philosophical and political premises of the ETI proposition, but contribute enormously to the evolution of a more acute consciousness of mankind's organizational backwardness and the conviction that, man at this critical stage, approaching the moment when he may meet a superior intelligence, need not perish but is capable to evolve toward a higher global order. The creation of an international study group, integrated by leading social scientists and that could call itself "Meaning and Impact of Search and Contact with Alien Intelligence," may be a first valid step in this direction.

Mass Media. The information industry of the Western World, especially the press, radio, television, and motion pictures, but also most of science fiction literature has so far given the issue of extraterrestrial intelligence a highly distorted and controversial treatment. By catering to the appetite for the exotic, mysterious, and shocking rather than to the need of sober understanding, it has helped nurse much of the ignorance and prejudice still distinguishing public opinion regarding the meaning of ETI.

Time has clearly arrived for a more responsible way of facing up to this intriguing scientific subject. Journalists, authors, and scriptwriters ought to give less importance to the lucrative commercial aspects of a conditioned market demand and realize the damage they are doing by diffusing crackpot theories and visions of extraterrestrial intelligent life, and by fanning on trivial and obscurant biases about the attitudes and intentions of extraterrestrials. And media executives, editors, and producers need to be more on the watch out for the scurrilous hodgepodge of imagery that unscrupulous and ill-informed minds heap onto their desks. More than ever, the public is entitled to well-balanced, critical, but above all, scientifically up-to-date information of the profound meaning of SETI and CETI, and I can't help thinking that newspapermen, writers, and producers with a sense of commitment to their large audiences might find the educative effort needed to correct the rampant vulgar and fraudulent speculations, and to foster realistic appraisals of the fantastic perspectives of contact not only a highly provocative, but also a surprisingly profitable, undertaking.

Education. As pedagogues and communication scientists know well, myth and prejudice, according to *Lévi Strauss,* "the fetters that bind us to the past," are characteristic traits of the adult population. To change a grown-up's beliefs on politics, religion, or science is often a terribly frustrating enterprise. This is why the message of search and contact with extraterrestrial intelligence must reach the young. But it must reach the young not only at the university level, where—as experiences at Ohio State University, Washington and Jefferson College, and Stanford University have shown—courses on ETI were extremely successful.[10] That there may be other life forms and intelligent life forms in the universe and that someday we will surely meet them and establish a thrilling and mutually beneficial relationship with them, is something that should be taught from the kindergarten and primary school level on through high school and college, to make sure that not only the recently graduated B.A. or Ph.D., but also all other youngsters, who dropped out of the school system at an earlier stage, start adulthood with a clear understanding of the scientific propositions on ETI.

But it would be wise to conceive curricula of SETI and CETI courses not only in terms of physical laws and astronomy, of radio waves and starships, and of the genetic code and microorganisms in space, though teaching these subjects, by using the ETI issue as a lead with a high motivational potential, will boost tremendously learning efficiency. Teachers should bring across the idea that extraterrestrial beings may be our peers not only technologically, but also socially and politically, and that contact with them may help unify our species and give us the chance to access the fabulous wealth of knowledge and expertise of other worlds. Broad efforts along this line would not only help jolting our stultified educational systems but contribute in transforming the proposition of ETI in a stirring human cause. If we succeed to implant this seed in the innermost of the young generations, that by reaching out to other intelligent beings in the universe, mankind's future and our lives will be immensely richer and more rewarding than our present situation, we also underwrite the best insurance for the unchecked progress and survival of our civilization.

But to recruit social sciences, mass media, and education systems in support of the deeper meaning of ETI will not be an easy task. It requires making people aware of the subtle linkage between search for extraterrestrial intelligence and search of solutions to mankind's global crisis; and only if this awareness becomes the flying banner of determined campaigns on the national and international level, leading to appropriate decisions in academic, corporate, but also government quarters, will the profound message of ETI blossom to full fruition. Much needs to be done in this respect. The efforts of the *Planetary Society* in the U.S., founded and directed by *Carl Sagan*, represent a most encouraging endeavor in this regard, that calls for similar initiatives in other parts of the world. It will be up to admirable initiatives as this to transform SETI into a conscious issue of average man and in a cause that may eventually move mountains.

At a recent conference about the future of Silicon Valley, the disjuncture "between what the technology can do and what society can do with the technology" received acute attention. Even more appalling, however, is the gap between the existing designs for a new global order, and what nations do with these designs. Evidently the pace of technological as well as organizational transformation accelerates if it is accompanied by change on the broad cultural front. The undaunted frontier spirit, imminent in the proposition of ETI, may play in this context a forceful and remarkably innovating role. With its future-oriented objectives, it may, furthermore, lend powerful support to the breakthrough of the new revolutionary paradigm of "anticipatory learning," advocated among others by a study group of the *Club of Rome,* in opposition to the stagnant and traditionally past-oriented educational systems still existing in most countries.[11] The cause of ETI may thus aid, in many important ways, to help create trend-swinging forces capable of superseding conservative immobilism and of mapping out and promoting a new global human project.

Skeptics may charge that this is utopian, that ETI cannot bring all that. But man today, surrounded by a sea of violence and hatred, facing scenarios of nuclear apocalypse, and diminished by the erosion of blind faith and ideology, is adrift and ready for a new and rational utopia. Following an obscure course of uncertainty, spaceship Earth needs a beacon of light, a tangible hope that not everything is forlorn for our civilization. Extraterrestrial intelligence, if it exists in our galaxy, may confront us with the biggest challenge to our hubris, but present us also the unique opportunity to raise above ourselves.

Epilogue

In reviewing the main lines of reasoning that have guided my writing in this book, I am astonished at how little needs refurbishment. Stalemate still has an ugly grimace and a firm grip on the world. Peace remains a romantic dream that mankind has learned to treat with profound skepticism and resignation. The international economic crisis continues to yawn at our doorstep, threatening to draw not only the reeling South but also the rebounding Western economies in the North into the whirlpool of universal bankruptcy. The North-South gap keeps growing and driving the two halves of the world irremediably apart, with rising inequalities, wants, and resentments building up pressures for new dangerous social turmoil and explosive bottlenecks. According to a UN report, released early in 1987, average per capita income of Third World countries decreased in 1986 in comparison with that of industrialized nations by 7.5 percent, and as was stressed by the International Monetary Fund in its report, "World Economic Perspectives," of April 1987, chances that countries as Argentina, Brazil, Mexico, and Venezuela will be able to pay their huge foreign debt must now be seen in a dimmer light than ever before. The possibility of Star Wars and nuclear winter, the spectre of a possible collapse of the international financial system, prospects of even more macabre famines than the one in Ethiopia in 1984–85 and the near certainty that international terrorism is likely to escalate to new climaxes, are some of the tantalizing future scenarios staring man in the face today. As we approach the finishing lap of this century, which has seen mankind plunge twice into lunatic genocide, bounce back with renewed confidence and verve, only to succumb again to vying animosities and holocaustic instincts, chances that we may soon advance toward a peaceful, more just and more rational world order seem remoter than ever.

Only in one respect has the global deadlock situation received a heartening blow. In the wake of the summit meetings of *President Reagan* and *Chairman Gorbachev*, the danger of an imminent nuclear exchange between the two superpowers has begun to recede, and if the third meeting between the world's top two leaders yields the results that high Soviet and American officials prognosticate, an arms reduction deal and a new dawn of detente and coexistence may be in the offing. Should these prospects materialize, it means that, at least for the time being, planet Earth may be spared the dreaded suicidal apocalypse, and that mankind will have another chance for reshuffling the deck of its global order.

Though still facing stiff opposition from much of the Party's and State's mossback bureaucracy, *Gorbachev's* reform movement under the slogans of *Glasnost* and *perestroika** seems to be making headway. The lift of the ban of the leading dissident *Andrei D. Sakharov* and his triumphant return to Moscow and ostensible rehabilitation, the introduction of direct voting procedures in factories, and the liberalization of the restrictive emigration procedures are some of the significant developments signaling that a more democratic wind has begun to blow in the Soviet Union. The Soviet overtures to withdraw the Red Army from Afghanistan and to reduce their military presence in some of the Eastern European countries, but above all, the Kremlin's apparent readiness to agree to the so-called "Zero-Solution" with regard to the middle and short-range nuclear systems in Europe, are major concessions that seem to indicate the awareness on the part of the new Soviet leadership that treading the ground of possible military confrontation with the West has not paid the expected dividends for world Communism. They highlight the Soviet leader's earnest determination to apply an iron broom to Russia's stultified economic system and his interest in—at least—a temporary reprieve of the simmering and costly conflict with the United States.

However, it is premature to overrate the significance of these developments. It must be taken with caution, whether the inveterate Party *nomenclatura,* the enshrined dogmas and its worldwide revolutionary commitments will permit *Gorbachev* to introduce the radical reform that are necessary to give Communism a human face, to call of the Communist bid for world conquest, and to help mankind to return to peace. The unexpected strong resistance to *Teng Hsiao-ping's* far-reaching liberalization and modernization program in China, which may yet be defeated by the ossified Marxist-Leninist dogmatists in this country as a "revisionist deviation," illustrate the portentous vagaries that may beset and doom even the most sincere attempts to recast Communism in a new mold. But there is also another side to this glittering coin of *Chairman Gorbachev's* aperture, often forgotten by much of the euphoric reaction to the perspective that maybe soon we will get hold of the tip of the mantle of peace and genuine coexistence. No degree of self-delusion can rule out the fact that the more democratic and more liberal Soviet Union will be a more formidable

*Openness and transformation.

foe of the U.S. and the entire West, economically and politically, and on the global scale.

With regard to the future role of the United States in the caldron of world politics, a few cautionary remarks also seem opportune. Though it has ridden under the leadership of *President Reagan* a remarkable wave of economic recovery and nationalist fervor and self-esteem, *Iranscam* has opened a wound that laid bare more than wishy-washy White House leadership and policy-making. It is as if *Iranscam* had let loose the ghosts that had been haunting the nation's most critical and forward-looking minds all along. The education system of the U.S. has slipped behind that of its top competitors. In spite of its vast R & D potential, Japan is threatening fast to become tomorrow's front-runner in science and technology. Year-long neglect in social welfare has created a backlog of unfulfilled necessities that rock the nation's conscience and undermine its efficacy. On the foreign front, too, Washington's hapless meddlings in Nicaragua emphasize its ambiguity in handling hostile revolutionary regimes in its own backyard and its precipitated infatuation with the "Zero-Solution," proposed by the Soviet Union for the European theater, overriding apprehensions that such a solution may be disadvantageous to Western Europe and NATO, are signs that the U.S. is still unprepared to come to terms with the new trends shaping world reality. The U.S. lacks not only a clear-cut response to *Mr. Gorbachev's* offensive, but it continues to lack a coherent political discourse, tailored to the world's most pressing needs, and the global far-reaching vision, from which the creation of a more rational and just world order may spring. In another order of magnitude, this holds true also for Western Europe. Reduced to a mere theater of rivaling Soviet-American military strategies and condemned to a mere handyman's role in world politics, and incapable of transforming the successful but porous economic community into something much more monumental, as the United Federation of Europe, its nations just continue living it up, torn between self-indulgence and fear, without a future design and content with short-term rewards and breathing spells.

It is therefore not a fortuitous phenomenon that the underlying current of pessimism that gained ground over the past forty years continues to be a pervasive trend, characterizing much of man's thinking today. It is a long string, stretching from the conviction that man is doomed to extinction, and that the rich North will never resolve the poor South's lot to the belief that peace is but a cruel chimera and technology the lure with which the devil paves the road to man's damnation. "Within the next ten years," prophesied the American computer expert *Joseph Weizenbaum,* "we will experience an information catastrophe, quite comparable, though on another scale, with an atomic catastrophe."[1] In their book, *The Fifth Generation,* authors *Edward A. Feigenbaum* and *Pamela McCurdock* also warn against the dangers of artificial intelligence and the progressive automatization of knowledge-acquisition procedures—their lack of transparency and the mental boredom and intellectual emasculation which allegedly will affect the masses of the post-industrial societies. From a different angle, *Mark Edmundson,* contributing editor to the fine journal "Channels of Communication," adverts to the rapid imposition of what *Marshall McLuhan* used to call a growing process of "perceptual numbing," in his own words the development of a "stubborn insensitivity to all but the most extreme experience in life."[2] These tendencies, it is held, may pose a greater threat to the intellectual virility and survival of the West than the drug and cult culture, corroding its entrails from within, and the Communist challenge, menacing it from without.

But mankind's real future may not be as bleak as painted by some of the pessimistic doomsdayists, who may yet earn the doubtful distinction of "associating wisdom with pessimism," as *John Kenneth Galbraith* once pointed out, and of having espoused the hopeless Luddites' cause at the close of the twentieth century. The law of the inexorability of historical development applies only to the past, not to the future. As the French utopian *Pierre Proughon* taught, mankind is never safe from surprises, only once they have become history, they are digested as the natural course of events. It is probably one of the ironies of our time, that its most arresting traits, the lurid East-West and North-South confrontation and the global crisis of stalemate, which is mankind's biggest calvary, are still grasped mainly as the upshot of sinister forces, rather than as the growing pains of a new world that is being born. Understood in this way, stalemate—for all its grisly and terrifying facets—is but the unresolved dilemma between the urgently needed global change and the barriers of the old structures and paradigms, no longer in touch with the demands of our time. But within the spiritual and ideological desert, threatening to engulf developed and underdeveloped nations alike, new change-oriented ideas and currents are budding, which may yet checkmate the vicious circles of stalemate.

One trend, particularly, works strongly in man's favor—our enormous preoccupation about the future. Never before, in all of our history, has man thought more and further about his future. Never has he been more obsessed with the idea what tomorrow will have in store for him, what future technology will mean to his life, what the world will look like at the turn of this century, and whether mankind has a future at all. Never has he projected his vision of future scientific and social scenarios further into other centuries than today. I think that this is an encouragingly healthy sign and that, in this context, search and contact with extraterrestrial intelligence have a significant role to play.

Mankind's future will most likely be inextricably intertwined with even more prodigious thrusts of science and technology. There is no way of curtailing constant progress in either of these fields. With the aid of artificial intelligence, man's cognitive capacities will be magnified manyfold and parallel to the accelerated evolution of the psychosocial component—at the expense of the purely biological one—his capability to understand and enjoy science, but also to mold the physical and social environment in the image of his innermost desires, will be greatly enhanced. Driven forward by the pace of the microelectronic and bio-genetic revolution and followed by new path-blazing breakthroughs in other fields, the twenty-first century will give rise to *Homo cientificus* on an unprecedented scale. If sanity prevails on the global scene, it may also lead to *Homo pacificus*.

Space exploration, in spite of the temporary U.S. setback due to the *Challenger* tragedy, is going strong, and will no doubt reach new heights in the decades ahead. In February 1986, the Soviets launched its space station *MIR*. Beginning in 1987, it is programmed to give shelter to ten scientists during ten months. U.S. efforts for orbiting a permanent manned space station by 1990 are progressing according to plan. NASA's Hubble space telescope is scheduled for orbiting in 1988. Ambitious space programs have also been initiated by the European Space Agency, Japan, China, and Canada. Even more important, the defrosting of Soviet-American relations is giving cooperative space undertakings a new lease on life. The *Hermes-MIR* linkup may become a reality already in 1988, and if the resumed talks between NASA and the Soviet Space Agency progress, the envisioned joint U.S. and Soviet Mars mission may be the big scientific feat of the midnineties. In its report, "The Next Giant Leap: An Agenda for International Cooperation," the United Nations Association of the U.S., which conducted a nationwide study in 1986 on the subject, came out strongly for a resumption of cooperative space ventures, advocating—as the study's Project Director *Ann Florini* stressed, "U.S. foreign policy interests and the long-term interests of its space program coincide."[3] These vigorous efforts and plans underline that man's step into space is irreversible and they lend support to the prospect that major international endeavors of space colonization will be undertaken by mankind in the twenty-first century. Search of extraterrestrial intelligence is bound to profit from this promising new cooperative thrust into space.

But the most monumental discoveries and feats of science and technology will not be enough to restore man's faith in the purpose of life. He needs a cause appealing to his Faustian adventure spirit, to his yearning to the beyond, but at the same time with a deep meaning for himself and the future of civilization. "Our only hope," said *Erich Fromm*, "is the energizing power which comes from a new vision." I think that the proposition on search and contact with ETI encapsulates much of this new vision.

The Western world is in desperate need of it. Its reserves of telluric creativity have not yet been exhausted, but its faith, that it will come out on top of present adversities, is no longer unshaken. Much of its original grand vision and political purposefulness have somehow been siphoned off from its marrow. The mobilizing strength of its basic political and ideological message is but a pale reflection of the glow that irradiated from it not too long ago. It is the task of the leading Western nations to summon up their intellectual and moral resources and to rally them in support of new universally valid goals, and to pursue them with the same passion and fierce determination with which their foes pursue the demise of the West and the implantation of a totalitarian but maybe more rational and just global order. If they fail, the political and ethical foundations on which their societies are built, and that have served mankind in its darkest hours, may collapse, weakened internally by a lack of purpose and vanquished by an external force with a more vital vision for leading the world out of its present plight and on to a higher stage along man's fantastic journey into the future.

APPENDIXES

Appendix A

First Soviet-American Conference on Communication with Extraterrestrial Intelligence (CETI)

Yerevan 1971

Conference conclusions:

1. The striking discoveries in recent years in the fields of astronomy, biology, computer science, and radio physics have transferred some of the problems of extraterrestrial civilization to a new realm of experiment and observation. For the first time in human history, it has become possible to make serious and detailed experimental investigations of this fundamental and important problem.

2. This problem may prove to be of profound significance for the future development of mankind. If extraterrestrial civilizations are ever discovered, the effect on human scientific and technological capabilities will be immense, and the discovery can positively influence the whole future of man. The practical and philosophical significance of a successful contact with an extraterrestrial civilization would be so enormous as to justify the expenditure of substantial efforts. The consequences of such a discovery would greatly add to the total of human knowledge.

3. The technological and scientific resources of our planet are already large enough to permit us to begin investigations directed towards the search for extraterrestrial intelligence. As a rule, such studies should provide important scientific results even when specific searches for extraterrestrial intelligence do not succeed. At present, these investigations can be carried out effectively in the various countries by their own scientific institutions. Even at this early stage, however, it would be useful to discuss and coordinate specific programs of research and to exchange scientific information. In the future, it would be desirable to combine the efforts of investigators of various countries to achieve the experimental and observational objectives. It seems to us appropriate that the search for extraterrestrial intelligence should be made by representatives of the whole of mankind.

4. Various modes of search for extraterrestrial intelligence were discussed in detail at the Conference. The realization of the most elaborate of these proposals would require considerable time and effort and an expenditure of funds comparable to the funds devoted to space and nuclear research. Useful searches can, however, also be initiated at a very modest scale.

5. The Conference participants consider highly valuable present and forthcoming space-vehicle experiments directed towards searching for life on other planets of our solar system. They recommend the continuation and strengthening of work in such areas as prebiological organic chemistry, searches for extrasolar planetary systems, and evolutionary biology, which bear sharply on the problem.

6. The Conference recommends the initiation of specific new investigations directed towards modes of search for signals. A list of some possible investigations is appended.

7. To coordinate national programs of search and to promote progress in this field, the Conference suggests the establishment by appropriate means of an international working group. For the time being, the following interim working group is proposed: F. Drake, U.S.A.; N.S. Kardashev, U.S.S.R.; P. Morrison, U.S.A.; B. Oliver, U.S.A.; R. Pešek, Czechoslovakia; C. Sagan, U.S.A.; I.S. Shklovsky, U.S.S.R.; G.M. Tovmasyan, U.S.S.R.; and V.S. Troitsky, U.S.S.R.

8. The Conference participants urge the full and open publication of research results on these problems, and as a step in this direction plan simultaneous publication in Russian and in English of the Proceedings of the present Conference.

9. The interim working group is instructed to consider, when necessary, convening broadly based or more specialized meetings of scientists working on CETI.

10. The Conference participants express their warm appreciation for the splendid hospitality extended to them by the Armenian Academy of Sciences.

(Signed)
The organizing Committees of the U.S. and U.S.S.R. delegations, for the Conference participants.

Appendix B

Notes for the *Voyager* Record

Greetings of *Kurt Waldheim, Secretary General of the United Nations* for the *Voyager* record:

As the Secretary General of the United Nations, an organization of 147 member states who represent almost all of the human inhabitants of the planet Earth, I send greetings on behalf of the people of our planet. We step out of our solar system into the universe seeking only peace and friendship, to teach if we are called upon, to be taught if we are fortunate. We know full well that our planet and all its inhabitants are but a small part of the immense universe that surrounds us and it is with humility and hope that we take this step.

Statement of *Jimmy Carter, President of the United States* for the *Voyager* record:

This *Voyager* spacecraft was constructed by the United States of America. We are a community of 240 million human beings among the more than four billion who inhabit the planet Earth. We human beings are still divided into nation-states, but these states are rapidly becoming a single global civilization.

We cast this message into the cosmos. It is likely to survive a billion years into our future, when our civilization is profoundly altered and the surface of the Earth may be vastly changed. Of the 200 billion stars in the Milky Way galaxy, some—perhaps many—may have inhabited planets and space-faring civilizations. If one such civilization intercepts *Voyager* and can understand these recorded contents, here is our message:

This is our present from a small distant world, a token of

our sounds, our science, our images, our music, our thoughts and our feelings. We are attempting to survive our time so we may live into yours. We hope someday, having solved the problems we face, to join a community of galactic civilizations. This record represents our hope and our determination, and our good will in a vast and awesome universe.

Appendix C

——————— The International SETI Petition ———————

The International SETI petition is printed here with an up-to-date list of signatories. The petition was prepared by Carl Sagan, with Planetary Society logistical support, and organized by Mary Maki.

THE HUMAN SPECIES is now able to communicate with other civilizations in space, if such exist. Using current radioastronomical technology, it is possible for us to receive signals from civilizations no more advanced than we are over a distance of at least many thousands of light years. The cost of a systematic international research effort, using existing radiotelescopes, is as low as a few million dollars per year for one or two decades. The program would be more than a million times more thorough than all previous searches, by all nations, put together. The results — whether positive or negative — would have profound implications for our view of our universe and ourselves.

WE BELIEVE such a coordinated search program is well-justified on its scientific merits. It will also have important subsidiary benefits for radioastronomy in general. It is a scientific activity that seems likely to garner substantial public support. In addition, because of the growing problem of radiofrequency interference by civilian and military transmitters, the search program will become more difficult the longer we wait. This is the time to begin.

IT HAS BEEN SUGGESTED that the apparent absence of a major reworking of the Galaxy by very advanced beings, or the apparent absence of extraterrestrial colonists in the solar system demonstrates that there are no extraterrestrial intelligent beings anywhere. At the very least, this argument depends on a major extrapolation from the circumstances on Earth, here and now. The radio search, on the other hand, assumes nothing about other civilizations that has not transpired in ours.

THE UNDERSIGNED are scientists from a variety of disciplines and nations who have considered the problem of extraterrestrial intelligence — some of us for more than 20 years. We represent a wide variety of opinion on the abundance of extraterrestrials, on the ease of establishing contact, and on the validity of arguments of the sort summarized in the first sentence of the previous paragraph. But we are unanimous in our conviction that the only significant test of the existence of extraterrestrial intelligence is an experimental one. No *a priori* arguments on this subject can be compelling or should be used as a substitute for an observational program. We urge the organization of a coordinated, worldwide and systematic search for extraterrestrial intelligence.

Sagan, Carl, David Duncan Professor of Astronomy and Space Sciences; Director, Laboratory for Planetary Studies; Center for Radiophysics and Space Research, Cornell University

Baltimore, David, Director, Whitehead Institute for Biomedical Research and Professor of Biology, Massachusetts Institute of Technology; Nobel Laureate in Physiology and Medicine

Berendzen, Richard, President, American University

Billingham, John, Chief, Extraterrestrial Research Division, NASA Ames Research Center

Burbidge, E. Margaret, Professor of Astronomy and Director, Center for Astrophysics and Space Sciences, University of California, San Diego; Former Director, Royal Greenwich Observatory, UK; President, American Association for the Advancement of Science; Past President, American Astronomical Society

Calvin, Melvin, University Professor of Chemistry and Former Director, Laboratory of Chemical Biodynamics, University of California at Berkeley; Nobel Laureate in Chemistry

Cameron, A. G. W., Professor of Astronomy, Harvard University; Former Chairman, Space Science Board, National Research Council, National Academy of Sciences

Chadha, M. S., Senior Researcher, Bhabha Atomic Research Centre, Bombay, India

Chandrasekhar, S., Morton D. Hull Distinguished Service Professor of Physics and Astrophysics, University of Chicago; National Medal of Science

Crick, Francis, Distinguished Research Professor, Salk Institute; Nobel Laureate in Physiology and Medicine

Dixon, Robert S., Assistant Director, Ohio State University Radio Observatory

Donahue, T. M., Professor of Atmospheric Sciences, University of Michigan; Chairman, Space Science Board, National Research Council, National Academy of Sciences

Drake, Frank D., Goldwin Smith Professor of Astronomy, Cornell University; Former Director, National Astronomy and Ionosphere Center

DuBridge, Lee A., President Emeritus, California Institute of Technology; Former Presidential Science Advisor

Dyson, Freeman J., Professor of Physics, Institute for Advanced Study, Princeton

Eigen, Manfred, Director, Section on Biochemical Kinetics, Max Planck Institute for Biophysical Chemistry, Gottingen, German Federal Republic; Nobel Laureate in Chemistry

Eisner, Thomas, Jacob Gould Schurman Professor of Biology, Cornell University

Elliot, James L., Associate Professor of Astronomy and Physics and Director, George R. Wallace Jr. Astrophysical Observatory, Massachusetts Institute of Technology

Field, George B., Senior Scientist, Smithsonian Astrophysical Observatory, and Professor of Astronomy, Harvard University; Former Director of the Harvard-Smithsonian Center for Astrophysics

Ginzburg, Vitaly L., Senior Staff Member, Lebedev Physical Institute, Moscow; Lenin Prize Laureate

Gold, Thomas, John L. Wetherill Professor of Astronomy and Former Director, Center for Radiophysics and Space Research, Cornell University

Goldberg, Leo, Former Director, Kitt Peak National Observatory; Past President, International Astronomical Union

Goldreich, Peter, Lee A. DuBridge Professor of Astrophysics and Planetary Physics, California Institute of Technology

Gott, J. Richard, III, Associate Professor of Astrophysics, Princeton University

Gould, Stephen Jay, Professor of Geology, and Alexander Agassiz Professor of Zoology, Harvard University

Hagfors, Tor, Director, National Astronomy and Ionosphere Center; Professor of Electrical Engineering, Cornell University; Former Professor of Electrical Engineering, University of Trondheim, Norway

Hawking, Stephen W., Lucasian Professor of Mathematics, Cambridge University, UK

Heeschen, David S., Senior Scientist and Former Director, National Radio Astronomy Observatory

Heidmann, Jean, Chief Astronomer, Paris Observatory

Herzberg, Gerhard, Distinguished Research Scientist, National Research Council of Canada; Nobel Laureate in Chemistry

Hesburgh, Rev. Theodore, President, University of Notre Dame

Horowitz, Paul, Professor of Physics, Harvard University

Hoyle, Fred, Former Plumian Professor of Astronomy and Experimental Philosophy and Former Director of the Institute of Astronomy, Cambridge University, UK

Jones, Eric M., Staff Member, Los Alamos Scientific Laboratory

Jugaku, Jun, Professor of Astronomy, University of Tokyo, Japan

Kardashev, N. S., Director, Samarkand Radio Observatory, Institute for Cosmic Research, Soviet Academy of Sciences, Moscow

Kellermann, Kenneth I., Senior Scientist, National Radio Astronomy Observatory

Klein, Michael J., Senior Scientist, Jet Propulsion Laboratory, NASA

Lee, Richard B., Professor of Anthropology, University of Toronto, Canada

Lindblad, Per-Olof, Professor of Astronomy and Director of the Stockholm Observatory, Stockholm, Sweden

MacLean, Paul D., Chief, Laboratory of Brain Evolution and Behavior, National Institute of Mental Health

Marov, Mikhail Ya., Department Chief, M. V. Keldysh Institute of Applied Mathematics, Soviet Academy of Sciences, Moscow; Professor of Planetary Physics, Moscow State University

Meselson, Matthew, Thomas Dudley Cabot Professor of the Natural Sciences and Professor of Biochemistry and Molecular Biology, Harvard University

Minsky, Marvin L., Donner Professor of Science and Former Director, Artificial Intelligence Laboratory, Massachusetts Institute of Technology

Morimoto, Masaki, Director, Nobeyama Radio Observatory, Tokyo, Japan

Morrison, Philip, Institute Professor, Massachusetts Institute of Technology

Murray, Bruce, Professor of Geological and Planetary Science, California Institute of Technology; Former Director, Jet Propulsion Laboratory, NASA

Newman, William I., Assistant Professor of Planetary Physics and Astronomy, University of California, Los Angeles

Oliver, Bernard M., Vice President for Research and Development (Ret.), Hewlett-Packard Corporation

Oort, J. H., Professor of Astronomy, Leiden University; Former Director, Leiden Observatory, Netherlands; Past President, International Astronomical Union

Öpik, Ernst J., Senior Scientist, Armagh Observatory, Armagh, Northern Ireland, UK

Orgel, Leslie E., Research Professor, The Salk Institute; Adjunct Professor of Chemistry, University of California, San Diego

Pacini, Franco, Director, Arcetri Observatory, Florence, Italy

Papagiannis, Michael D., Chairman, Department of Astronomy, Boston University; President, Commission on the Search for Extraterrestrial Life, International Astronomical Union

Pauling, Linus, Former Chairman, Division of Chemistry and Chemical Engineering, California Institute of Technology; Chairman of the Board, Linus Pauling Institute of Science and Medicine; Nobel Laureate in Chemistry; Nobel Laureate in Peace; International Lenin Peace Prize Laureate; National Medal of Science

Pešek, Rudolf, Professor of Fluid Mechanics Emeritus, Faculty of Mechanical Engineering, Czech Technical University, Prague; Chairman, Commission on Astronautics, Czechoslovak Academy of Sciences

Pickering, W. H., Professor of Electrical Engineering Emeritus, California Institute of Technology; Former Director, Jet Propulsion Laboratory, NASA; National Medal of Science

Ponnamperuma, Cyril, Professor of Chemistry and Director, Laboratory of Chemical Evolution, University of Maryland

Purcell, Edward M., Gade University Professor Emeritus, Harvard University; Nobel Laureate in Physics; National Medal of Science

Raup, David M., Chairman, Department of Geophysical Sciences, University of Chicago; Former Dean of Science, Field Museum of Natural History; Former Professor of Geology and Paleontology, University of Rochester

Reber, Grote, Inventor of the Radiotelescope, Tasmania

Rees, Martin J., Plumian Professor of Astronomy and Director of the Institute of Astronomy, Cambridge University, Cambridge, UK

Russell, Dale A., Chief, Paleobiology Division, National Museums of Canada, Ottawa, Canada

Sagdeev, Roald Z., Director, Institute for Cosmic Research, Soviet Academy of Sciences, Moscow

Shannon, Claude E., Former Research Mathematician, Bell Laboratories; Donner Professor of Science Emeritus, Massachusetts Institute of Technology; National Medal of Science

Shklovskii, I. S., Chairman, Astrophysics Division, Institute for Cosmic Research, Soviet Academy of Sciences, Moscow

Tarter, Jill, Research Astronomer, University of California, Berkeley

Thomas, Lewis, Chancellor, Memorial Sloan-Kettering Cancer Center; Professor of Medicine, Cornell University Medical School

Thorne, Kip S., William R. Kenan Jr. Professor and Professor of Theoretical Physics, California Institute of Technology

Troitsky, V. S., Scientific Director, Radiophysics Research Institute, Gorky, USSR

von Hoerner, Sebastian, Senior Staff Member, National Radio Astronomy Observatory

Wilson, Edward O., Baird Professor of Science and Professor of Biology, Harvard University; National Medal of Science

Zuckerman, Benjamin, Professor of Astronomy, University of Maryland

[Affiliations are for identification purposes only]

SETI – An International Petition

Sir, The human species is now able to communicate with other civilisations in space, if such exist. Using existing radio-astronomical technology, it is possible for us to receive signals from civilisations no more advanced than we are over a distance of at least many thousands of light years. The cost of a systematic international research effort, using existing radiotelescopes, is as low as a few million dollars per year for one or two decades. The programme would be more than a million times more thorough than all previous searches, by all nations, put together. The results – whether positive or negative – would have profound implications for our view of our Universe and ourselves.

We believe such a coordinated search programme is well-justified on its scientific merits. It will also have important subsidiary benefits for radioastronomy in general. It is a scientific activity that seems likely to garner substantial public support. In addition, because of the growing problem of radiofrequency interference by civilian and military transmitters, the search programme will become more difficult the longer we wait. This is the time to begin.

It has been suggested that the apparent absence of a major reworking of the Galaxy by very advanced beings, or the apparent absence of extraterrestrial colonists in the Solar System demonstrates that there are no extraterrestrial intelligent beings anywhere. At the very least, this argument depends on a major extrapolation from the circumstances on Earth, here and now. The radio search, on the other hand, assumes nothing about other civilisations that has not transpired in ours.

The undersigned are scientists from a variety of disciplines and nations who have considered the problem of extraterrestrial intelligence – some of us for more than 20 years. We represent a wide variety of opinion on the abundance of extraterrestrials, on the ease of establishing contact, and on the validity of arguments of the sort summarised in the first sentence of the previous paragraph. But we are unanimous in our conviction that the only significant test of the existence of extraterrestrial intelligence is an experimental one. No *a priori* arguments on this subject can be compelling or should be used as a substitute for an observational programme. We urge the organisation of a coordinated, worldwide and systematic search for extraterrestrial intelligence.

Carl Sagan, Cornell University.
David Baltimore, Massachusetts Institute of Technology.
Richard Berendzen, American University,
John Billingham, NASA Ames Research Center.
Melvin Calvin, University of California at Berkeley.
A. G. W. Cameron, Harvard University.
M. S. Chadha, Bhabha Atomic Research Centre, Bombay, India.
S. Chandrasekhar, University of Chicago.
Francis Crick, Salk Institute.
Robert S. Dixon, Ohio State University.
T. M. Donahue, University of Michigan.
Frank D. Drake, Cornell University.
Lee A. DuBridge, California Institute of Technology.
Freeman J. Dyson, Institute for Advanced Study.
Manfred Eigen, Max Planck Institut Göttingen, German Federal Republic.
Thomas Eisner, Cornell University.
James Elliott, Massachusetts Institute of Technology.
George B. Field, Harvard University.
Vitaly L. Ginzburg, Lebedev Physical Institute, Moscow.
Thomas Gold, Cornell University.
Leo Goldberg, Kitt Peak National Observatory.
Peter Goldreich, California Institute of Technology.
J. Richard Gott, III, Princeton University.
Stephen Jay Gould, Harvard University.
Tor Hagfors, National Astronomy and Ionosphere Center.
Stephen W. Hawking, Cambridge University, Cambridge, UK.
David S. Heeschen, National Radio Astronomy Observatory.
Jean Heidmann, University of Paris.
Gerhard Herzberg, National Research Council of Canada.
Rev. Theodore Hesburgh, University of Notre Dame.
Paul Horowitz, Harvard University.
Fred Hoyle, Cambridge University, Cambridge, UK.
Eric M. Jones, Los Alamos Scientific Laboratory.
Jun Jugaku, University of Tokyo.
N. S. Kardashev, Institute for Cosmic Research, Soviet Academy of Sciences Moscow.
Kenneth I. Kellerman. National Radio Astronomy Observatory,
Richard B. Lee, University of Toronto.
Per-Olof Lindblad, Stockholm Observatory.
Paul D. MacLean. National Institute for Mental Health.
Matthew Meselson, Harvard University.
Marvin L. Minsky, Massachusetts Institute of Technology.
Masaki Morimoto, Nobeyama Radio Observatory, Tokyo, Japan.
Philip Morrison, Massachusetts Institute of Technology.
Bruce Murray, California Institute of Technology.
William I. Newman, University of California, Los Angeles.
Bernard. M. Oliver, Hewlett-Packard Corporation.
J. H. Oort, Leiden University.
Ernst J. Öpik, Armagh Observatory.
Leslie Orgel. Salk Institute.
Franco Pacini, Arcetri Observatory.
Michael D. Papagiannis, Boston University.
Linus Pauling, Linus Pauling Institute for Science and Medicine.
Rudolf Pesek, Czechoslovak Academy of Sciences.
W. H. Pickering, California Institute of Technology.
Cyril Ponnamperuma, University of Maryland.
Edward M. Purcell, Harvard University.
David M. Raup, University of Chicago.
Grote Reber, Tasmania.
Martin J. Rees, Institute of Astronomy, Cambridge University, Cambridge, UK.
Dale A. Russell, National Museums of Canada.
Roald Z. Sagdeev, Institute for Cosmic Research, Soviet Academy of Sciences, Moscow.
I. S. Shklovskii, Institute for Cosmic Research, Soviet Academy of Sciences, Moscow.
Jill C. Tarter, University of California, Berkeley.
Lewis, Thomas, Memorial Sloan-Kettering Cancer Center.
Kip S. Thorne, California Institute of Technology
Sebastian von Hoerner, National Radio Astronomy Observatory.
Edward O. Wilson. Harvard University.
Benjamin Zuckerman, University of Maryland.

Medical Considerations for Manned Interstellar Flight

Sir, I found the paper "Medical Considerations for Manned Interstellar Flight" by J. R. Murphy in the November 1981 issue to be of very great interest, particularly in view of my own contribution in the previous issue [1]. The basic difference in the two papers is that Murphy presumes that prolonged life in space will impose very serious problems

Appendix D

Objectives of IAU Commission 51:

1. The search for planets in other solar systems;
2. The evolution of planets and their ability to sustain life over cosmic periods;
3. The search for biologically relevant interstellar molecules and the study of their formation;
4. The search for radio signals, intentional or unintentional, of extraterrestrial origin;
5. The search for different manifestations of other advanced civilizations;
6. The spectroscopic detection of biological activity of primitive forms of life in other stars;
7. The coordination and promotion of all these activities at the international level, and the collaboration with other international organizations (astronautical, biological, chemical, et cetera) that share with our Commission common interests in these objectives.

Notes

CHAPTER 1
OUR PUZZLING STALEMATE

1. Arthur Koestler, *Der Mensch Irrläufer der Evolution* (Bern and Munich: Scherz Verlag, 1978), 123.

2. Richard Berendzen, ed., *Life Beyond Earth and the Mind of Man* (Washington, D.C.: NASA, 1973), 17f.

CHAPTER 2
LOST FAITH IN POLITICS AND POLITICIANS

1. Peter F. Drucker, *The Age of Discontinuity* (New York: Harper and Row, 1968), 213.

2. Thomas W. Madron, "Political Parties in the 1980s," *The Futurist*, Vol. XIII, No. 6 (December 1979), 467f.

3. Raymond Williams, *The Year 2000* (New York: Pantheon Books, 1983), 172.

4. Robert L. Heilbroner, *An Inquiry into the Human Prospect* (Canada: N.W. Norton, Inc., 1974), 14–15.

5. Winston S. Churchill, *Memoirs of the Second World War* (New York: Bonanza Books, 1978), 886.

6. Erich Fromm, *Sein und Haben* (Munich: Deutscher Taschenbuch Verlag, 1979), 21.

7. Quoted by W.W. Rostow, *Getting from Here to There* (New York: McGraw-Hill Company, 1978), 97.

8. George F. Kennan, *The Nuclear Delusion* (New York: Pantheon Books, 1982), 145.

CHAPTER 3
THE EAST-WEST BOTTLENECK

1. Richard Nixon, *The Real War* (New York: Warner Books, 1980), 18.

2. Quoted by Richard J. Barnet, *The Giants, Russia and America* (New York: Simon and Schuster, 1977), 124.

3. Konrad Lorenz, *On Aggression* (New York: Bantam, 1977), 90.

4. George F. Kennan, *The Nuclear Delusion, Soviet-American Relations in the Atomic Age* (New York: Pantheon Books, 1983), 231.

5. Jean Francois Revel, *How Democracies Perish* (New York: Doubleday, 1984), 240.

6. Robert F. Ellsworth, *Eine Bewertung des weltweiten militärischen Kräfteverhältnisses* (Bonn: Europa-Archiv, 1984), 171. (The aforementioned contribution has been published in issue 6/1984 of *Europa-Archiv,* (c) Verlag Für Internationale Politik GMBH, Bonn).

7. Richard Nixon, *Real Peace* (Boston: Little, Brown, and Company, 1983), 16.

8. Thomas B. Larson, *Soviet-American Rivalry* (New York: W.W. Norton and Company, 1978), 279.

9. Aspen Institute for Humanistic Studies, *Managing East-West Conflict* (Washington, D.C.: Aspen Institute for Humanistic Studies, 1983), 13.

10. Alexander Solzhenitsyn, "Misconceptions about Russia Are a Threat to America," *Foreign Affairs,* Vol. 38, No. 4 (Spring 1980), 833.

CHAPTER 4
THE DEEPENING NORTH-SOUTH GAP

1. *Report of the United Nations Conference on Science and Technology for Development* (New York: United Nations, 1979), 9.

2. *Global 2000 Study* (Washington, D.C.: US Government, 1980), 39.

3. Jean van der Terk, Carl Haube, and Elaine Murphy, "A New Look at the Population Problem," *The Futurist,* Vol. XIV (April 1980), 39.

4. *Global 2000 Study,* 13.

5. *World Economic Outlook 1984* (Washington, D.C.: International Monetary Fund, 1984), 44.

6. "Entwicklungsländer in der Liquiditätsfalle," *Wirtschaftswoche,* No. 8 (February 17, 1984), 2.

7. *World Development Report,* published for the World Bank (Oxford: Oxford University Press, 1984), 32–33.

8. David Horovitz, "The North and the South," *International Development Report* (1977).

9. Mahbub ul Haq, *The Poverty Curtain* (New York: Columbia University Press, 1976), 35, 74.

10. International Press Service cable, from Geneva, dated June 15, 1984.

11. Richard R. Fagan, "Equity in the South in the Context of North-South Relations," from *Rich and Poor Nations in the World Economy, 1980s Project/Council on Foreign Relations* (New York: McGraw-Hill, 1980), 180.

12. Robert Ramsay, "UNCTAD's Failures: The Rich Get Richer," *International Organization,* Vol. 38, No. 2 (Spring 1984), 393f.

13. "Der Bundestag übt Kritik an den Waffenkäufen der Entwicklungsländer," *Frankfurter Allgemeine Zeitung* (March 6, 1982).

14. *North-South: A Program for Survival: The Report of the Independent Commission on International Development Issues under the Chairmanship of Willy Brandt* (Cambridge: MIT Press, 1980), 230f.

15. "Does aid of the rich make the poor only more dependent?" *Kölner Stadtanzeiger* (September 30, 1984).

16. Mahbub ul Haq, *Poverty Curtain,* 180.

CHAPTER 5
THE TRAGIC SPIRAL OF THE ARMAMENT RACE

1. "A Talk with Louis Harris," *Bulletin of the Atomic Scientists* (August/September 1982), 4. (Reprinted by permission of the *Bulletin of Atomic Scientists,* a magazine of science and world affairs, Copyright (c) 1982 by the Educational Foundation For Nuclear Science, Chicago, IL 60637).

2. Mark A. Harwell, *Nuclear Winter* (New York: Springer Verlag, 1984), 165.

3. Frank Barnaby, "World Arsenals in 1980," *Bulletin of the Atomic Scientists* (September 1980), 12. (Reprinted by permission. Copyright (c) 1980).

4. Henry W. Kendall, "Second Strike," *Bulletin of the Atomic Scientists* (September 1979), 32. (Reprinted by permission. Copyright (c) 1979).

5. Office of Technology Assessment, Congress of the U.S., *The Effects of Nuclear War* (London: Croom Helm, 1980), 27.

6. Jonathan Schell, *The Fate of the Earth* (New York: Alfred A. Knopf, 1982), 54f.

7. Office of Technology Assessment, *Nuclear War,* 94.

8. *Effects of Nuclear War on Health and Health Services* (Geneva: World Health Organization, 1984), 23.

9. Paul Ehrlich, Carl Sagan, et al., *The Cold and the Dark: The World after Nuclear War* (New York: W.W. Norton and Company, 1984), 18.

10. Ibid, 47f.

11. Harwell, *Nuclear Winter,* 6–7.

12. Barnaby, *World Arsenals,* 13.

13. Eugenia V. Osgood, "Euromissiles: Historical and Political Realities," *Bulletin of the Atomic Scientists* (December 1983), 18.

14. Robert Jastrow, "Reagan vs. the Scientists: Why the President Is Right about Missile Defense," *Commentary* (January 1984), 31.

15. Richard Halloran, "Pentagon Draws up First Strategy for Fighting a Long Nuclear War," *New York Times* (May 30, 1982).

16. "Victory in Space Is Possible," *Der Spiegel* (November 12, 1984), 145f.

17. William M. Arkin, "The Drift toward First Strike," *Bulletin of the Atomic Scientists* (January 1985), 5. (Reprinted by permission. Copyright (c) 1985).

18. Pierre E. Trudeau, "World Leaders Must Reassert Primacy," *Bulletin of the Atomic Scientists* (February 1985), 5. (Reprinted by permission. Copyright (c) 1985).

19. Quoted by Erhard Eppler in *Wege aus der Gefahr* (Hamburg: Rowohlt, 1981), 86.

20. Ibid. 79.

CHAPTER 6
WORLD PEACE: A DREAM GONE WRONG

1. Richard J. Barnet, *The Giants, Russia and America* (New York: Simon and Schuster, 1977), 124.

2. Lester Brown, *World without Borders* (New York: Vintage Books, 1973),

304. Ervin Laszlo, et al., *Goals for Mankind: A Report to the Club of Rome on the New Horizons of Global Community* (New York: E.P. Dutton, 1977), 146.

3. Jack Donnelly, "Recent Trends in UN Human Right Activities: Description and Polemics," *International Organization,* No. 4 (Autumn 1981), 654.

4. Richard Nixon, *The Real War* (New York: Warner Books, 1980), 69.

5. Herman Kahn, *On Thermonuclear War* (Princeton: Princeton University Press, 1961), 627.

6. Arnold Toynbee, *A Study of History* (Oxford: Oxford University Press and Thames and Hudson Ltd., 1972), 317f.

7. Alvin Toffler, *The Third Wave* (New York: William Morrow and Company, 1980), 343.

8. Carl F. von Weizsäcker, *Der bedrohte Friede* (Munich: Carl Hanser Verlag, 1981), 209.

9. Herman Kahn, *The Next 200 Years* (New York: William Morrow and Company, 1976), 207.

10. George F. Kennan, *The Cloud of Danger* (Boston: Little, Brown, and Company, 1977), 185.

11. Quoted by Richard A. Falk, *This Endangered Planet* (New York: Vintage Books, 1971), 39.

12. Gerd von Haßler, *Welt ohne Notausgang* (Bern: Scherz Verlag, 1984), 34.

CHAPTER 7
THE QUICKSANDS OF REVOLUTION

1. International Press Service report from Montevideo on March 18, 1985.

2. *Informe de la Comisión Nacional Bipartita sobre Centroamerica* (Colombia: Carbajal, S.S., 1984), 5.

3. Cynthia McClintock, "Why Peasants Rebel: The Case of Peru's Sendero Luminoso," *World Politics,* No. 1 (October 1984), 59.

4. Associated Press report from Bogota on September 22, 1983.

5. Carl F. von Weizsäcker, *Der bedrohte Friede* (Munich: Carl Hanser Verlag, 1981), 566.

6. *Der Spiegel* (June 22, 1981), 125.

7. Interview reproduced by the weekly resumee of the Cuban official newspaper *Gramma* on February 24, 1985.

8. Quoted by Carlos Rangel in *Del buen salvaje al buen revolucionario* (Caracas: Monte Avila Editores, C.A., 1977), 352.

9. Roger Garaudy, *La Alternativa* (Madrid: Editorial Cuadernos para el Dialogo, S.A., 1975), 250.

CHAPTER 8
SCIENCE AND TECHNOLOGY: BETWEEN GLORIFICATION AND DAMNATION

1. Quoted by Nigel Calder in *Technopolis,* German edition (Düsseldorf: Econ Verlag, 1971), 349.

2. James Everett Katz, *Presidential Politics and Science Policy* (New York: Praeger Publishers, 1978), 9.

3. Philip H. Abelson, "Biotechnology: An Overview," *Science* (February 11, 1983), Vol. 219, No. 4585, 612. (Reprinted by permission of the American Association For the Advancement of Science).

4. Steve Sinclair, "Dialogues on America's Future—Computers Will Replace Human Intelligence," Address to the U.S. Congress on March 29, 1984.

5. E. Cooper, "Science in Underdeveloped Countries," in *The Next 200 Years* by Herman Kahn (Paris: OECD, 1968), 162.

6. Arthur C. Clarke, *Profiles of the Future* (New York: Harper and Row, 1973), IX.

7. Friedrich Cramer, *Fortschritt durch Verzicht* (Munich: Nymphenburger Verlagshandlung, 1975), 182.

8. W.W. Rostow, *Getting from Here to There* (New York: McGraw-Hill, 1978), 151.

9. Dieter Balkhausen, *Die Dritte Industrielle Revolution* (Düsseldorf: Econ Verlag, 1978), 190.

10. J.D. Bernal, *The World, the Flesh, and the Devil* (Bloomington: Indiana University Press, 1969), 77.

11. Erhard Eppler, *Wege aus der Gefahr* (Hamburg: Rowohlt Verlag, 1981), 180.

12. Clarke, *Profiles,* 116–117.

13. Cramer, *Fortschritt,* 266.

14. Jacques Monod, *Zufall und Notwendigkeit* (Munich: Deutscher Taschenbuchverlag, 1983), 149.

15. Fritjof Capra, *The Turning Point* (New York: Simon and Schuster, 1982), 375.

16. Carl Friedrich von Weizsäcker, *Der bedrohte Friede* (Munich: Carl Hanser Verlag, 1981), p. 559.

17. Erich Fromm, *Haben und Sein* (Munich: Deutscher Taschenbuch Verlag, 1979), 167.

18. Carl Sagan, *Brocca's Brain* (New York: Random House, 1974), 277.

CHAPTER 9
THE ESCAPIST URGE

1. Alvin Toffler, *Future Shock* (New York: Random House, 1970), 450.

2. George Gallup, Jr., *Adventures in Immortality* (New York: McGraw-Hill, 1982), 3, 183, 187.

3. "On the Wave of the Occult," *Der Spiegel* (November 30, 1981), 232.

4. Martin Gardner, *Science—Good, Bad, and Bogus* (Buffalo: Prometheus Books, 1981), 244.

5. Larry Kusche, *The Bermuda Triangle Solved* (New York: Warner Books, 1975).

6. *The Skeptical Inquirer* (Fall 1978), 6; (Fall 1981), 42.

7. "Fauler Zauber," *Der Spiegel* (October 25, 1982), 262.

8. John Skow, "The Genesis of Equal Time," *Science* (December 1981), 54. (Reprinted by permission from December issue of *Science* '81 by the American Association For the Advancement of Science).

9. "Beträchtliche Spanne," *Der Spiegel* (November 29, 1982), 101.

10. Amaury de Riencourt, *The Eye of Shiva: Eastern Mysticism and Science* (New York: William Morrow and Co., 1981), 65.

11. Eugene Herrigel, *The Method of Zen* (New York: Vintage Books, 1974), 55.

12. William Sims Bainbridge and Rodney Stark, "Superstitions: Old and New," *The Skeptical Inquirer* (Summer 1980), 24.

13. "Million-Pound Saucer Hunt," *The Skeptical Inquirer* (Fall 1978), 9; "Betty through the Looking Glass," *The Skeptical Inquirer* (Winter 1979–80), 880.

14. Erich von Däniken, *Habe ich mich geirrt?* (Munich: C. Bertelsmann, 1985), 211ff.

15. Walt Michalsky, "The Masquerade of Fundamentalism," *The Humanist* (July–August 1981), 18.

16. David A. Conway, "Jastrow and Genesis," *Free Inquiry* (Winter 1981), 32f.

17. George Gallup, Jr., *Immortality,* 183f.

18. W.S. Bainbridge and R. Stark, "Superstitions," 24.

19. Quoted by Kendrick Frazier, "News and Comments," *The Skeptical Inquirer* (Spring 1982), 6.

20. "A Nation at Risk: The Imperative for Educational Reforms," *A Report to the Nation and the Secretary of Education* (Washington D.C.: The National Commission of Excellence in Education, April 1983), 5.

CHAPTER 10
THE GLOBAL DEADLOCK

1. Erhard Eppler, *Wege aus der Gefahr* (Hamburg: Rowohlt, 1981), 17.

2. Arnold Brecht, *Kann die Demokratie Überleben?* (Stuttgart: Deutsche Verlags-Anstalt, 1978), 103.

3. Michael J. Crozier, Samuel P. Huntington, and Joji Watanuki, "The Crisis of Democracy," *Report on the Governability of Democracies to the Trilateral Commission* 1975, 3.

4. Daniel Bell, *The Cultural Contradictions of Capitalism* (New York: Basic Books, 1978), 84.

5. Svetozar Stojanovich, *In Search of Democracy in Socialism* (Buffalo: Prometheus Books, 1981), 56.

6. Leszed Kolakowski, "In Poland Marxism Is a Dead Thing," *Der Spiegel* (September 8, 1980), 129.

7. Regis Debray, *Les Empires contre l'Europe* (Paris: Gallimard, 1985), 303.

8. Folke Dovring, *Soviet-American Rivalry* (New York: W.W. Norton and Co., 1978), 272.

9. Rudolf Augstein, "Papier auf das man schreibt," *Der Spiegel,* (November 25, 1985), 140.

10. Gerd von Hassler, *Welt ohne Notausgang* (Bern: Scherz Verlag, 1984), 99.

11. George F. Kennan, *The Cloud of Danger* (Boston: Little, Brown, and Co., 1977), 31.

12. Richard J. Barnet, *The Giants* (New York: Simon and Schuster, 1977), 84.

13. Horst E. Richter, *Alle redeten vom Frieden* (Hamburg: Rowohlt, 1981), 58.

14. Teng Hsiao-ping, "Kapitalistische Dinge stärken neuerdings den Sozialismus," *Entwicklungspolitick* (November 1985), 348.

15. Alexander Solzhenitsyn, "Misconceptions about Russia," *Foreign Affairs*, Vol. 58, No. 4 (1980).

16. Wolfgang Seiffert, *Kann der Ostblock Überleben?* (Bergisch Gladbach: Gustav Lübbe Verlag, 1983), 116.

17. Richard Falk, *This Endangered Planet* (New York: Vintage Books, 1971), 422, 427

18. Rene Dubos, *Celebrations of Life* (New York: McGraw-Hill, 1981), 251.

CHAPTER 11
THE BASIC PROPOSITION

1. Herman Kahn, *The Coming Boom* (New York: Simon and Schuster, 1982), 228.

2. Mikhail S. Gorbachev, *A Time for Peace* (New York: Richardson and Steirman, 1985), 266.

3. Carl Sagan and Thornton Page, *UFOs: A Scientific Debate* (Ithaca: Cornell University Press, 1972), 272.

4. Walter Sullivan, *We Are Not Alone* (New York: McGraw-Hill, 1964), 288.

5. Carl Sagan, *The Cosmic Connection* (London: Coronet Books, 1975), 33.

6. Isaac Asimov, *Extraterrestrial Civilizations* (New York: Crown Publishers, 1979), 271–2.

7. Giuseppe Cocconi, "SETI Is Coming of Age," in *The Search of Extraterrestrial Life: Recent Developments*, edited by Michael D. Papagiannis (Boston: D. Reidel Publishing, 1985), 518.

8. Jack Catran, *Is There Intelligent Life on Earth?* (Sherman Oaks, Ca.: Lidiraven Books, 1980), 159.

CHAPTER 12
SEARCH FOR EXTRATERRESTRIAL INTELLIGENCE:
AN IMPRESSIVE RECORD

1. N.S. Kardashev, "Communication and Extraterrestrial Civilization," in *Extraterrestrial Civilization* (Washington, D.C.: NASA and National Science Foundation, 1965).

2. Quoted by Frank Drake, "The Foundations of the *Voyager* Record," in *Murmurs of Earth*, by Carl Sagan and others (New York: Random House, 1978), 65.

3. Philip Morrison, John Billingham, and John Wolfe, eds., *The Search for Extraterrestrial Intelligence, SETI* (Washington, D.C.: NASA, 1977), 1.

4. "Possibility of Intelligent Life Elsewhere in the Universe," Report prepared for the Committee on Science and Technology, U.S. House of Representatives, by the Science Policy Research Division (Washington, D.C.: U.S. Government Printing Office, October 1977).

5. "Seventeenth Report by the International Telecommunications Union on Telecommunication and the Peaceful Uses of Outer Space," A/AC. 105/213, December 22, 1977.

6. Quoted by Frank Drake, "SETI Tallinin 1981, Communication with Extraterrestrial Intelligence," *Cosmic Search* (first half of 1982), 4.

7. NASA SETI Program, *The Search for Extraterrestrial Intelligence* (Moffett Field, Ca.: SETI Program Office, NASA, Ames Research Center, 1984).

8. Michael D. Papagiannis, "Search for Extraterrestrial Life, A New Commission of the International Astronomical Union," *Astrosearch* (March/April 1983), 4.

9. Michael D. Papagiannis, ed., *The Search for Extraterrestrial Life: Recent Developments* (Boston: IAAU, D. Reidel Publishing Company, 1985), 518.

10. Bernard M. Oliver, "A More Eclectic Approach," in *Extraterrestrial Life,* edited by Michael D. Papagiannis, 352.

11. "SETI—An International Petition," *JBIS* (March 1983), 143.

12. *Astronomy and Astrophysics for the 1980s, Volume 1: Report of the Astronomy Survey Committee* (Washington, D.C.: National Academy Press, 1982), 150.

CHAPTER 13
PROBABILITIES THAT EXTRATERRESTRIAL LIFE AND INTELLIGENCE EXIST: THE STAND OF SCIENCE

1. Isaac Asimov, *Interstellar Contact* (Chicago: Regnery Company, 1974), 223f.

2. Carl Sagan, *Cosmos* (New York: Random House, 1980), 296.

3. J.H. Wolfe, et al., "SETI—Search for Extraterrestrial Intelligence: Plans and Rationale," in *Life in the Universe,* edited by John Billingham (Cambridge: MIT Press, 1982), 392.

4. Michael D. Papagiannis, ed., *The Search for Extraterrestrial Life: Recent Developments* (Boston: D. Reidel Publishing Co., 1985), 29.

5. Billingham, *Life in the Universe,* 392.

6. J. John Sepkoski, Jr., "Some Implications of Mass Extinction for the Evolution of Complex Life," in *Recent Developments,* edited by Michael D. Papagiannis, 230.

7. Cyril Ponnamperuma, "Primordial Organic Chemistry," in *Extraterrestrials—Where Are They?,* edited by Michael D. Papagiannis (New York: Pergamon Press, 1982), 95.

8. Isaac Asimov, *Interstellar Conflict,* 189.

9. Carl Sagan, *The Dragons of Eden* (New York: Random House, 1977), 16.

10. Papagiannis, *Recent Developments,* 180.

11. Isaac Asimov, *Interstellar Conflict,* 189.

12. V.A. Ambartsumian in *Communication with Extraterrestrial Intelligence*, edited by Carl Sagan (Cambridge: MIT Press, 1973), 7.

13. Papagiannis, *Recent Developments*, 546.

CHAPTER 14
NUMBER OF EXTRATERRESTRIAL CIVILIZATIONS: ESTIMATES AND DEBATE

1. Carl Sagan, ed., *Communication with Extraterrestrial Intelligence, CETI* (Cambridge: MIT Press, 1973), 166.

2. Isaac Asimov, *Extraterrestrial Civilizations* (New York: Crown Publishers, 1979), 91.

3. James R. Wertz, "The Human Analogy and the Evolution of Extraterrestrial Civilizations," *JBIS*, Vol. 29 (1976), 101.

4. Alan Bond and Anthony R. Martin, "A Comparative Estimate of the Number of Habitable Planets in the Galaxy," *JBIS* (March 1980), 101.

5. Sebastian von Hoerner, "Aspects of Interstellar Communication," *Astronautica Acta*, Vol. 18 (New York: Pergamon Press, 1973), 44f.

6. Ronald Bracewell, *The Galactic Club: Intelligent Life in Outer Space* (San Francisco: W.H. Freeman and Company, 1974), 72.

7. Sebastian von Hoerner, "The Likelihood of Interstellar Colonization and the Absence of Its Evidence," in *Extraterrestrials: Where Are They?*, edited by Michael H. Hart and Ben Zuckerman (New York: Pergamon Press, 1982), 162.

8. Frank J. Tipler, "The Most Advanced Civilization in the Galaxy is Ours," *Mercury* (January/February 1982), 5-6.

9. Michael H. Hart, "Atmospheric Evolution, the Drake Equation, and DNA: Sparse Life in an Infinite Universe," in *Extraterrestrials*, edited by M.H. Hart and Ben Zuckerman, 162.

10. M.H. Hart and Ben Zuckerman, *Extraterrestrials*, 161.

11. J.S. Shklovskii and Carl Sagan, *Intelligent Life in the Universe* (Oakland: Holden-Day, 1966), 450.

12. Carl Sagan, *Cosmos* (New York: Random House, 1980), 310.

13. John H. Wolfe, "On the Question of Interstellar Travel," in *The Search for Extraterrestrial Life: Recent Developments*, edited by Michael D. Papagiannis (Boston: D. Reidel Publishing Company, 1985), 452.

14. Nikolai S. Kardashev, "The Inevitability and the Possible Structures of Supercivilizations," in *Recent Developments*, ed. by Michael D. Papagiannis 497.

15. Isaac Asimov, *Extraterrestrial Civilizations*, 160.

16. Michael D. Papagiannis, "Colonies in the Asteroid Belt, or a Missing Term in the Drake Equation," in *Extraterrestrials*, ed. by M.H. Hart and Ben Zuckerman, 82.

17. Papagiannis, *Recent Developments*, 439.

18. J. Billingham, B. M. Oliver, and J.H. Wolfe, "A Review of the Theory on Interstellar Communication," in *Communication with Extraterrestrial Intelligence*, ed. by J. Billingham and R. Pešek (New York: Pergamon Press, 1979), 55.

19. Michael D. Papagiannis, "Search for Extraterrestrial Life," *Astrosearch,* Vol. 1, No. 1 (March/April 1983); and Papagiannis *Recent Developments,* 546.

20. V.A. Firsoff, *Life among the Stars* (London: Allan Wingate, 1974), 113.

21. Quoted in *Contact with Extraterrestrial Intelligence, CETI,* edited by Carl Sagan (Cambridge: MIT Press, 1973), 346–7.

CHAPTER 15
ETI: ITS ADVANCED TECHNOLOGICAL AND ORGANIZATIONAL ORDER

1. J.S. Shklovskii and Carl Sagan, *Intelligent Life in the Universe* (San Francisco: Holden-Day, 1966), 394; and Carl Sagan, *The Cosmic Connection* (London: Coronet Books, 1975), 222.

2. Freeman Dyson, *Disturbing the Universe* (New York: Harper and Row, 1979), 213.

3. Nikolai S. Kardashev, "Strategy for the Search for Extraterrestrial Intelligence," in *Communication with Extraterrestrial Intelligence,* edited by John Billingham and Rudolf Pešek (Oxford: Pergamon Press, 1979), 37.

4. Carl Sagan, *Cosmos* (New York: Random House, 1980), 311.

5. Isaac Asimov, *Extraterrestrial Civilizations* (New York: Crown Publishers, 1979), 174.

6. V. Sevastyanov, A. Ursul, and Y. Shkolenko, *The Universe and Civilization* (Moscow: Progress Publishers, 1981), 213.

7. Nigel Calder, *Spaceships of the Mind* (London: British Broadcasting Corporation, 1978), 131.

8. Zdenek Kopal, *Man and His Universe* (London: Rupert Hart-Davies, 1972), 306–7.

9. Michael D. Papagiannis, *The Search for Extraterrestrial Life: Recent Developments* (Boston: D. Reidel Publishing Company, 1985), 508.

10. Freeman Dyson, letter to *Scientific American,* 1964, quoted by Walter Sullivan in *We Are Not Alone* (New York: McGraw-Hill, 1964), 131.

11. Arthur C. Clarke, *The Promise of Space* (New York: Harper and Row, 1968), 306.

12. Robert Nisbet, *History of the Idea of Progress* (New York: Basic Books, 1980).

13. Ervin Laszlo, et al., *Goals of Mankind* (New York: New American Library, 1978), 364.

14. Alvin Toffler, *The Third Wave* (New York: William Morrow and Company, 1980), 28.

15. Charles L. Seeger, "Fermi Question, Fermi Paradox: One Hit, One Out," in *Recent Developments,* edited by Michael D. Papagiannis, 488.

16. Nikolai S. Kardashev, "Strategy. . . ," 36.

CHAPTER 16
CONTACT WITH ETI: WHAT IT WOULD MEAN TO US

1. Eugen Sänger, *Raumfahrt—technische Überwindung des Krieges* (Munich: Ernesto Grassi, 1958), 110.

2. Donald N. Michael, editor, *Proposed Studies on the Implications of Peaceful Space Activities* (Washington, D.C.: prepared for NASA by the Brookings Institution, U.S. Government Printing Office, 1961), 216.

3. William S. Bainbridge, *The Spaceflight Revolution* (New York: John Wiley and Sons, 1976), 18.

4. Joseph Royce, "Consciousness and the Cosmos," in *Extraterrestrial Intelligence: The First Encounter,* edited by James L. Christian (Buffalo: Prometheus Books, 1976), 180.

5. Michael A.G. Michaud, "The Consequences of Contact," *AIAA Student Journal* (Winter 1977), 20f.

6. Carl Sagan and J.S. Shklovskii, *Intelligent Life in the Universe* (San Francisco: Holden-Day, 1966), 441.

7. T.A. Heppenheimer, *Colonies of Space* (Harrisburg: Stackpole Books, 1977), 175.

8. Bernard M. Oliver, "Technical Considerations in Interstellar Communications," in *Interstellar Communication: Scientific Perspectives,* by Cyril Ponnamperuma and A. G. Cameron (Boston: Houghton Mifflin, 1974), 141.

9. Isaac Asimov, *Extraterrestrial Civilizations* (New York: Crown Publishers, 1979), 273.

10. Reinhard Breuer, *Kontakt mit den Sternen* (Frankfurt: Umschau Verlag Breidenstein KG, 1978), 222.

11. Ibid., 285.

12. Ernst Fasan, *Relations with Alien Intelligences* (Berlin: Berlin Verlag, 1970), 66.

13. Ralph Chapman, ed., *The World in Space: A Survey of Space Activities and Issues* (New York: Prentice Hall, prepared for United Nations, 1982), 184.

14. Carl Sagan, *The Dragons of Eden* (New York: Random House, 1977), 233.

15. Duncan Lunan, *Interstellar Contact* (Chicago: Henry Regnery Company, 1974), 200.

16. Carl Sagan, *Communication with Extraterrestrial Intelligence* (Cambridge: MIT Press, 1973), 345.

17. Gerald O'Neill, *The High Frontier* (New York: William Morrow and Company, 1977), 168.

CHAPTER 17
THE TASK BEFORE US

1. Carl Sagan, *The Cosmic Connection* (London: Coronet Books, 1975), 179.

2. George Wald, quoted by Ronald L. Bracewell in *Intelligent Life in Outer Space* (San Francisco: W.H. Freeman and Company, 1974), 75.

3. Philip Morrison, John Billingham, and John Wolfe, eds., *The Search for Extraterrestrial Intelligence, SETI* (Washington, D.C.: NASA, 1977), 33.

4. Ashley Montagu, quoted in *Life beyond Earth and the Mind of Man,* edited by Richard Berendson (Washington, D.C.: NASA, 1963), 24.

5. "Fifth International Review Meeting of Communication with Extraterrestrial Intelligencies,"*JBIS* (March 1977), 119.

6. "Messages to Extraterrestrial Civilizations," *UN Committee on the Peaceful Uses of Outer Space,* A/AC, 105/206 (October 18, 1977), 8.

7. Duncan Lunan, *Interstellar Contact* (Chicago: Henry Regnery Company, 1974), 209.

8. Michael A.G. Michaud, "Toward A Grand Strategy for the Species," *Astrosearch,* (November/December 1983), 8.

9. "Moscow's Program Takes Off," *Time* (March 31, 1986), 2. (Copyright 1986, Time, Inc. All rights reserved. Reprinted by permission from *Time*).

10. "SETI Popular in Colleges," *Cosmic Search* (Fall 1979), 39.

11. James W. Botkin, *No Limits to Learning: Bridging the Human Gap* (Berlin: Wilhelm Goldmann Verlag, 1979).

EPILOGUE

1. Joseph Weizenbaum, Interview, *Frankfurter Rundschau* (November 30, 1983).

2. Mark Edmundson, "McLuhan: It's All Going According to Marshall's Plan," *Channels of Communications* (May/June 1984), 49.

3. Ann Florini, "The Next Giant Leap: Space Exploration as Foreign Policy," *The Planetary Report* (January/February 1987), 12–3.

Index

A

Abelson, Philip H., 89
Adenauer, Konrad, 9, 13
Aleksandrov, Vladimir, 46
Allard, Paul, 30
Allende, Salvador, 78
Andropov, Yuri, 51, 54
Anson, Jay, 103
Aquino, Corazon, 25
Arafat, Yasser, 73, 74
Arkin, William M., 53
Arrhenius, Svante A., 155
Asimov, Isaac, 115, 137, 150,
    160, 165, 175, 187, 191
Augstein, Rudolf, 124

B

Baader-Meinhof Group, 72, 76
Bainbridge, W. S., 106, 184
Baker Plan, 37
Balkhausen, Dieter, 93, 94, 96
Bandung Conference, 33
Barnet, Richard, 128
Bauer, Lord, 37
Belaunde, Terry F., 80
Bell, Coral, 62
Bell, Daniel, 120
Bell, T. H., 115
Benitez, Leopoldo, 62
Berg, David, 107, 112
Berger, Peter, 122
Berlitz, Charles, 105, 202
Bernal, J. D., 94
Betancourt, Belisario, 73
Bethe, Hans, 56
Bhagwan Shree Rajnesh, 108,
    111, 115
Billingham, John, 151
Birks, John W., 46

Black Panthers, 72
Bohlen, Charles, 24
Bohm, David, 114
Bond, Alan, 160
Bracewell, Ronald, 161, 166,
    175
Brandt, Willy, 20, 33
Brecht, Arnold, 12, 118
Breuer, Reinhard, 191
Brezhnev, Leonid, 18, 23, 121
Brown, Harold, 62
Brown, Lester, 61
Brzezinski, Zbigniew, 125
Bundy, McGeorge, 70

C

Capitalism, 121, 122
Capra, Fritjof, 97, 114
Carter, Jimmy, 11, 17, 25,
    128, 144
Castro, Fidel, 8, 37, 74, 77, 78,
    83
Catron, Jack, 139
Chamberlain, Neville, 10, 11
Chargaff, Erwin, 92
Chernenko, Konstantin, 13
Churchill, Winston, 9, 14
Clarke, Arthur C., 92, 95
    101, 176, 178
Cleaver, Eldridge, 75, 77
Clements, R. D., 113
Club of Rome, 126, 214
Cocconi, Giuseppe, 140, 141
Commission—Search for
    Extraterrestrial Life, 146
Commoners, Barry, 86
Communism (Soviet), 121, 122
Comte, Auguste, 169
Condon Report, 109

Constellation Ophiuchus, 186
Contact with Extraterrestrial
    Intelligence, 5, 139, 210,
    211, 231
Convergence theory, 125, 126
Cooper, E., 90
Copperman, Paul, 115
Corea, Germani, 34
Cox, Harvey, 110
Cramer, Friedrich, 92, 95, 96
Crick Francis, 155, 169, 190
Crozier, Michael, 119
Cruitzen, Michael, 46
Cuban Revolution, 77, 83
Curcio, Renato, 77

D
Dadzie, K.K.S., 34
Däniken, Erich von, 110, 163
Davies, Angela, 75
Debrais, Regis, 76, 79, 123
Declaration of Quito, 33
De Gaulle, Charles, 9, 87
Detente, 20, 22, 23
Difurth, von, 106
Djilas, Milovan, 124
Dixon, Robert, 147
Dole, Stephan, 151, 160
Donnelly, Jack, 62
Dovring, Folke, 123
Dozier, James L. 77
Drake, Frank, 140, 142, 150,
    172, 189
Dror, Yehezkel, 10, 119
Drucker, Peter F., 6
Dubos René, 132
Dulles, John Foster, 19
Dutschke, Rudi, 75
Dyson, Freeman, 157, 171,
    175

E
Edmundson, Mark, 218
Ehrlich, Paul, 46, 47
Eigen, Manfred, 166
Einstein, Albert, 189, 190

Eisely, Loren, 185
Eisenhower, Dwight, 16, 91
Ellsworth, Robert F., 23
Eppler, Erhard, 58, 118
Extraterrestrial Intelligence
    (ETI), 139
    future tasks, 199f
    meaning of contact, 183f
    number of extraterrestrial
        civilizations, 159, 160
    probability of existence,
        150, 151f
    search for, 5, 139, 143, 147,
        148, 206, 207
    with an advanced
        technological and
        organizational order, 5,
        139, 210, 211f

F
Fagan, Richard A., 34
Falk, Richard, 132
Fannon, Frantz, 75
Fasan, Ernst, 194
Feigenbaum, Edward, 218
Fermi, Enrico, 161
Fermi Paradox, 163, 165, 166,
    167
Fine, Terrence, 158
Firsoff, V. A., 168
Flammerion, Camille, xi
Florini, Ann, 219
Fromm, Erich, 10, 99, 220
Franco, Francisco, 13

G
Gaddafi, Muammar el, 8, 11
Gairy, Eric Mathew, 205
Galbraith, John Kenneth, 218
Gallup, George, Jr., 114
Garaudy, Roger, 84
Gardner, Martin, 105
Geller, Uri, 105
Geneva Summit Meeting, 26,
    131, 135, 210, 215
Ghandi, Indira, 73

Global Report 2000, 30, 31
Gorbachev, Mikhail, 13, 21, 23,
    26, 54, 69, 121, 123, 125,
    130, 136, 210, 215, 217
Gray, Colin S., 53
Green Bank Conference, 141
Guevara, Ché, 74, 76, 83

H
Hackett, John, 17, 42
Haldane, John, B. S. 153
Hall, Cordell, 62
Hamilton, William, 187
Harris, Louis, 42
Harris, Marvin, 25
Haq, Mahbub ul, 32
Hart, Michael, 162, 165, 166
Harwell, Mark A., 43, 47
Hassler, Gerd von, 70, 127
Hearst, Patty, 77
Heilbroner, Robert L., 8, 120
Herrigel, Eugen, 108
Hitler, Adolf, 10, 65
Hoerner, Sebastian von, 160,
    161, 166, 196
Horovitz, David, 32
Horovitz, Paul, 146
Hoyle, Fred, 187
Hubbarb, Lafayette, R., 107
Hurd, Paul, 115
Huxley, Aldous, preface
Hynek, Allen, 109

I
IAU Symposium 112
    Search for Extraterrestrial
        Life and Recent
        Developments, 146
Interfutures, 30

J
Jaruzelski, Wojciech, 25, 128
Jarvik, Robert, 92
Jastrow, Robert, 114
Jessup, Phillip, 65
Jones, Jim, 107
Jung, Carl, 195

K
Kahn, Herman, 14, 41, 64, 67,
    90, 94, 136
Kardashev, N. S., 141, 165,
    171, 182
Katz, James E., 86
Kelsen, Hans, 65
Kendall, Henry W., 44, 45
Kennan, George F., 14, 21, 68,
    127
Kennedy, Donald, 48
Kennedy, J. F., 11, 20, 33
Khomeini, Ruhollah, 8, 74
Kirkpatrick, Jeane, 62
Kissinger, Henry, 20, 22
Kissinger Commission, 73
Kistiakowsky, George B., 53
Koestler, Arthur, 4, 10, 11
Kohn, Bendit, 75
Kolakowski, Leszek, 122
Kopal, Zdenek, 175
Krauss, Melvyn, 37
Khrushchev, Nikita, 11, 23,
    121, 123
Kurtz, Paul, 105
Kusche, Larry, 105
Kuznets, Simon, 39

L
Larson, Thomas, 25
Lasswell, Harold, 10
Liberal Democracy, 118, 119
Lima, Turcios, 76
Lorenz, Konrad, 21
Lunan, Duncan, 208

M
Madron, Thomas W., 6
Majaraj Ji, 108
Malcolm X, 74
Malraux, André, 69, 127
Mao Tse-tung, 11
Marcos, Ferdinand, 35
Martin, Anthony R., 160
MacArthur, Douglas, 16
McClintock, Cynthia, 73

McCown, Delos B., 112
McCurdock, Pamela, 218
McLuhan, Marshall, 180, 218
McNamara, Robert, 29, 49
Michaud, A. G. Michael, 187,
   209
Miller, Stanley Lloyd, 153,
   165
Mitterand, Francois, 38, 87
Monod, Jacques, 96, 100
Montagu, Ashley, 175, 203
Morgenthau, Hans, 8, 65
Morrison, Philip, 140, 150,
   195, 200
Murchison Meteorite, 155

N
New International Economic
   Order (NIEO), 32, 33, 62
New International
   Information Order (NIIO),
   62
Newman, William, 164
Nisbet, Robert, 177
Nixon, Richard, 17, 24, 42, 62,
   210
Nixon/Mao Tse-tung Summit
   Meeting, 32
North-South Brandt Report,
   29, 30, 33, 36, 37, 54, 126
Nuclear Winter, 46, 47
Nüey, Lecomte du, 165
Nyerere, Julius, 35

O
Oliver, Bernard M., 147, 187,
   190
O'Neill, Gerald, 132, 175, 196
Oparin, Aleksandr, 153
Oppenheimer, Robert, 57
Orgel, Leslie E., 155
Ortega, Daniel, 141, 183
OZMA One, 141, 183
OZMA Two, 143

P
Palme, Olof, 73
Palme Commission, 54
Palmer, Patrick, 143
Papagiannis, Michael D., 146,
   147, 150, 157, 158, 166, 167,
   175
Pascal, Blaise, 10
Pešek, Rudolf, 183
Pichat, Jean Bourgeois, 89
Pinochet, Augusto, 80
Pioneer 10 and *11*, 142
Planetary Society, 213
Ponnamperuma, Cyril, 150,
   153, 154, 165
Price, Don, 88
Project Cyclops, 142
Project Daedalus, 143
Project META, 147
Project Sentinel, 147
Project Serendip, 147
Proxmire, William, 145, 201
Puente Uceda, 76

R
Rabinowitz, Eugen, 95
Ramsay, Robert, 34, 35
Randi, James, 105
Randolph, Bernard, 53
Reagan, Ronald, 8, 11, 13, 21,
   22, 25, 26, 52, 53, 69, 125,
   128, 136, 149, 210, 215, 217
Roberts, Walter Orr, 86

S
Sakharov, Andrei, 26, 124,
   130, 216
Sandinista Revoluton, 81, 82
Savasta, Antonio, 77
Schell, Jonathan, 43, 44
Schumacher, E. F., 40, 90, 96
Schumpeter, Joseph, 93
Schmidt, Helmut, 12, 17
Seeger, Charles J., 159, 166,
   181

Seiffert, Wolfgang, 132
Sendič, Raul, 72
Sepkovski, J. John, Jr., 154
Servan-Schreiber, Jean-
    Jacques, 98
Sevastyanov, V., 175
Shcharansky, Anatoli, 177
Shklovskii, Josip, 141, 145,
    166, 170
Shkolenko, Yu, 175
Sinclair, Steve, 90
Sinsheimer, Robert L., 196
Stockholm International
    Peace Research Institute
    (SIPRI), 48
Sirica, John J., 96
Skow, John, 106
Smith, Harlan, 167
Smith, Samantha, 54
Somoza, Tacho, 7, 74, 81
Sorokin, Pitrim, 125, 126
Soviet Centralized Planned
    Economy, 123, 124
Space Telescope, 152
Space Treaty, 204, 205
Spall, N. J., 200
Spielberg, Steven, 147, 174,
    200
Stalin, Josef, 9, 10, 12, 16
Stojanovich, Svetozar, 121
Stark, R., 106
Strategic Defense Initiative
    (S.D.I.), 11, 20, 26, 51, 129
Strauss, Levi, 212
Sullivan, Walter, 137, 200
Sun Moon, 107, 133
Szilard, Leo, 95

T
Teng Hsiao-ping, 13, 122, 129,
    216
Thatcher, Margaret, 11
Thorsson, Inga, 49
Timbergen, Jan, 125
Tipler, Frank J., 162

Tito, J., 11, 13
Tocqueville, Alexis de, 16
Toffler, Alvin, 66, 102, 106,
    118, 180
Tomin, Julius, 127
Torres, Camilo, 74, 76
Toure, Sekou, 11
Toynbee, Arnold, 64, 177
Trotzkii, Josip, 145
Trudeau, Pierre, 55
Truman, Harry, 11, 18
Tsiolkovski, Konstantin, 166,
    180
Tucker, Robert W., 32

U
Ulbricht, Walter, 11, 13
United Nations, 60f
    and ETI, 203f
U.N. Committee on the
    Peaceful Use of Outer
    Space, 204
U.N. Committee for Search
    and Contact with ETI, 206,
    207
Ursul, A., 175
U.S. Defense Department, 45

V
Vallee, Jacques, 109
Velasco Alvarado, Juan, 80
Verne, Jules, xi
Verschuur, G., 143
Voyager I and *II*, 143, 154

W
Wald, George, 5, 201
Waldheim, Kurt, 144
Walford, Roy, 89
Wallace, Michael D., 53
Watson, James D., 190
Weber, Max, 114
Weber, Renée, 114
Weinberger, Caspar, 51, 53
Weizenbaum, Joseph, 218

Weizsäcker, Carl Friedrich
    von, 21, 24, 65, 67, 75, 99,
    126
Wells, H. G., 174
Wertz, James A., 160
Williams, Raymond, 7
Wilson, Harald, 87
Wolfe, John H., 164

Z
Zero option, 54
Zoo Hypothesis, 163
Zuckerman, Benjamin, 143,
    161